SpringerBriefs in Applied Sciences
and Technology

SpringerBriefs in Computational Mechanics

Series Editors

Holm Altenbach 🆔, Faculty of Mechanical Engineering,
Otto-von-Guericke-Universität Magdeburg, Magdeburg, Sachsen-Anhalt, Germany

Lucas F. M. da Silva, Department of Mechanical Engineering, Faculty of
Engineering, University of Porto, Porto, Portugal

Andreas Öchsner, Faculty of Mechanical Engineering, Esslingen University of
Applied Sciences, Esslingen, Germany

These SpringerBriefs publish concise summaries of cutting-edge research and practical applications on any subject of computational fluid dynamics, computational solid and structural mechanics, as well as multiphysics.

SpringerBriefs in Computational Mechanics are devoted to the publication of fundamentals and applications within the different classical engineering disciplines as well as in interdisciplinary fields that recently emerged between these areas.

More information about this subseries at https://link.springer.com/bookseries/8886

Reza Beygi · Eduardo Marques ·
Lucas F. M. da Silva

Computational Concepts in Simulation of Welding Processes

Reza Beygi
INEGI—Campus da FEUP
Porto, Portugal

Eduardo Marques
INEGI—Campus da FEUP
Porto, Portugal

Lucas F. M. da Silva
Institute of Science and Innovation
in Mechanical and Industrial Engineering
Porto, Portugal

ISSN 2191-530X   ISSN 2191-5318 (electronic)
SpringerBriefs in Applied Sciences and Technology
ISSN 2191-5342   ISSN 2191-5350 (electronic)
SpringerBriefs in Computational Mechanics
ISBN 978-3-030-97909-6   ISBN 978-3-030-97910-2 (eBook)
https://doi.org/10.1007/978-3-030-97910-2

This Springer imprint is published by the registered company Springer Nature Switzerland AG
The registered company address is: Gewerbestrasse 11, 6330 Cham, Switzerland

Contents

Chapter 1
Introduction

1.1 The Importance of Simulating Welding Processes

Modelling and simulation of welding processes is a powerful engineering tool that is gaining importance in industrial applications. If implemented correctly, it allows to precisely predict how a welding process will take place and what will be its final result, determining its geometry, resultant microstructure and even the mechanical performance. However, the simulation of welding processes is still not as widespread as is the case for other production processes. For example, the use of modelling in forming processes is much more common, where the relative simplicity of the process contrasts with the complex nature of welding. In general, welding modelling requires more data, more variables and thus significantly more computational time. Furthermore, in most industrial applications, welding is now part of a large process chain and optimizing all these processes is a very time-consuming activity. Therefore, several assumptions and simplifications must be applied to reduce simulation time and ensure the practicality of a simulation-based approach.

Before any simplification is made, one must precisely determine the main objective of the optimization process, since in practice only one or a very limited number of objectives can be effectively determined in a simplified model. For example, a structure may experience a critical distortion during welding which must be limited and controlled (a transverse or longitudinal shrinkage or bending) and therefore one can simplify the model so that it only outputs the value of this critical distortion. In some cases, one may even restrict calculations to a particular region or divert more computational time towards that region because the changes of the objective in other regions are negligible. Another possible simplification relies on local control of the mesh and model complexity. For example, during welding, the area around the weld seam experiences higher temperature and stress gradients. Therefore, a finer mesh and a more precise mechanical model can be adopted for this area, allowing to take non-linearities into account while ensuring lower computational costs.

© The Author(s), under exclusive license to Springer Nature Switzerland AG 2022
R. Beygi et al., *Computational Concepts in Simulation of Welding Processes*,
SpringerBriefs in Computational Mechanics,
https://doi.org/10.1007/978-3-030-97910-2_1

Fig. 1.1 Simulation
approaches in welding

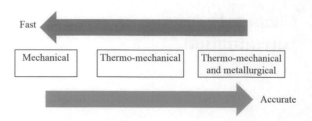

Several different goals can be pursued with a welding simulation, such as the determination of temperature distribution, distortion, residual stress and possible phase transformation in the material. It is important to optimize the welding process in order to achieve the desired shape and performance of the component. Simulation of the welding process offers the possibility to achieve this goal with lower costs and less time expenditure but the final accuracy of the simulation depends on whether all realistic conditions are taken into account. Of course, introducing all realistic physical phenomena in the model, such as metallurgical changes during the simulation, greatly increases the accuracy of the simulation but exponentially increases the necessary calculation time (Fig. 1.1), which means that a balance must always be established between the running time and the accuracy of the simulation. All simulation approaches used for modelling welding procedures aim to achieve both goals with minimal compromises in both time and accuracy. Much of this book is concerned with the strategies that can be used to achieve both goals.

Thermal and mechanical analyses are two important components of welding simulations, which yield temperature and residual stress/distortion, respectively. The following subsections describe why it is important to calculate temperature through simulation. In addition, the importance of simulation for the determination of residual stress and distortion is clarified.

1.2   Heat and Temperature Associated to Welding

The temperature–time curve at certain spatial locations and the temperature streamlines at certain times during welding provide important information about the process. The most important features of the time–temperature curve are the maximum temperature and the cooling rate. The most important characteristic of the temperature distribution is the temperature gradient. A schematic representation of the temperature–time curve is shown in Fig. 1.2, which shows the variation in peak temperature and cooling rate at different points around the fusion zone. Development of residual stress and distortion is strongly influenced by the temperature history and the material properties (both physical and mechanical properties). Furthermore, depending on the type of alloy, several phase changes can occur during welding, which can greatly influence the material properties and thus the residual stresses and distortion.

Fig. 1.2 Schematic of temperature–time history curves in some pints around the fusion zone

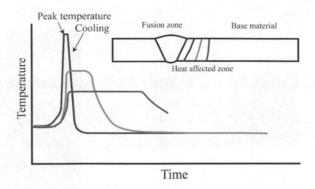

1.2.1   Experimental Methods to Measure Temperature

Experimental methods can be used to measure the temperature but have important limitations in its capabilities to determine spatial and temporal measurements. One of the most practical experimental methods used to measure temperature during welding is the use of an infrared thermometer or camera [1]. While the thermometer provides the temperature measurement of a specific area on the target surface, an infrared thermography camera works on a much larger region and can measure the temperature distribution [2]. However, the main limitation of these experimental techniques is that they only measure the temperature of the surface of the material. Thermocouples are another important tool for temperature measurement that can be placed in multiple locations to precisely determine the temperature there. The most significant limitation of this method is that thermocouples often cannot be placed in the most critical areas of interest. For example, in friction stir welding (FSW) one often wishes to determine the temperature as close as possible to the weld bead but the rotating tool in the stir zone will inevitably destroy the thermocouple [3]. The same thing happens in the fusion zone of fusion welding processes. Furthermore, the areas around the welding zone are often very critical and the temperature gradient there is extremely steep, especially in welding processes with an intense heat source such as laser welding. This makes the use of experimental techniques even more difficult, since it is not possible to measure the temperature correctly or reliably within such small intervals. This problem becomes more important when welding materials that undergo a critical phase transition in a low temperature range and thus in a narrow dimensional range. One example is the complete or partial dissolution of precipitates during laser welding of a Mg-Gd-Y-Zr alloy, which only occurs in a narrow region of the heat affected zone (HAZ) as shown in Fig. 1.3 [4]. Other examples are carbide precipitation in some alloy steels [5], grain coarsening in P91 steel [5], and martensite formation in high strength steel [6].

(a)   (b)

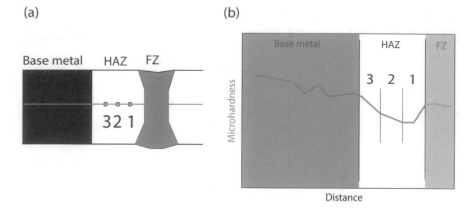

Fig. 1.3 a Schematic representation of different regions in weld, FZ (Fusion zone), heat affected zone (HAZ), and base metal. **b** Hardness profile in different region of a joint performed by laser welding on a Mg-Gd-Y-Zr alloy. Adopted from [4] with permission from Elsevier

1.2.2   Analytical Methods

Before the advancement of computers enabled any kind of numerical simulation, analytical models were extensively developed and widely used to calculate the temperature distributions in welds, both spatially and temporally. Diverse analytical formulae were thus proposed to calculate the evolution of temperature during welding. The most famous of these is the classical Rosenthal equation [7]. The temperature distribution around a moving heat source in x direction (welding direction) is obtained from [8]:

$$T = T_0 + \frac{\lambda P}{2\pi k r} \exp\left[-\frac{V(r + \xi)}{2\alpha}\right] \qquad (1.1)$$

where T_0 is the temperature at locations far from the top surface (initial temperature), k is the thermal conductivity, P is the power of the heat source, r is the distance from the heat source, V is the scanning velocity, and α is the thermal diffusivity. The heat source moves along the x-axis and the moving coordinate of x − Vt is replaced by ξ (the radial distance from the heat source).

When deriving the Rosenthal equation, it is assumed that the heat loss due to convection and radiation is negligible, especially with a high heat input, and therefore the accuracy of this model is not acceptable in these cases. It is reported that the use of the Rosenthal model leads to an overestimation of the weld pool boundary due to this effect [8]. In addition, this equation does not take into account the natural change in key material properties with temperature. Thermal conductivity (λ) and volumetric heat capacity (ρC) both change as the temperature increases. In some cases (for example in austenitic stainless steel), the correlation of these two parameters with temperature (T) is linear as follows [9]:

Fig. 1.4 Schematic representation of a comparison between the analytical models developed for prediction of temperature by considering the material properties correction, without it and the one obtained by FEM simulation

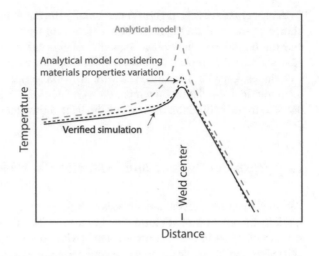

$$\lambda(T) = \lambda_0(1 + mT) \tag{1.2}$$

where λ_0 is the thermal conductivity at $T = 0$ and m is a coefficient. A correction was made to take this dependency into account in the analytical model and to increase its accuracy [9]. As shown in Fig. 1.4, the analytical model result comes close to the result of the numerical simulation by accounting for this variation of the material properties. It is important to take into consideration that the simplifications used to derive these analytical solutions restricts their applications to specific cases. For example, analytical models mostly assume an infinite or semi-infinite workpiece and do not consider the heat loss due to convection or radiation. Several physical mechanisms such as the flow of molten metal, the latent heat effect, the heat transfer through the melt and the formation of molten metal in the electrode are not taken into account in analytical formulas, which further reduces the accuracy [10].

1.2.3 Numerical Methods

The use of numerical simulations for calculating the temperature distribution is a powerful process that does not exhibit most of the limitations that exist in analytical models. In fact, the variation of material properties with regards to temperature, finite dimensions and complex geometries can all be easily implemented in the numerical simulation. Furthermore, such simulations also offer the possibility of calculating spatial and temporal temperature distributions with high accuracy, all without having to directly and experimentally measure the temperature. However, the use of numerical simulation does not fully avoid the need for experimental measurements since a few predetermined experiments are still essential to calibrate and validate the numerical model.

A numerical model suitable for calculating the temperature depends on the proper choice of the heat source model, the calibration technique, the material properties and the boundary conditions. Chapter 2 of this book explains several models for heat sources that represent different welding processes as well as the validation techniques. Chapter 3 describes how the material properties influence the results of the numerical simulation for temperature calculation. Finally, a section of Chap. 5 is devoted to the calibration methods of the heat source models.

1.3   Residual Stresses and Distortion in Welding

Welding residual stresses are stresses which are produced due to temporally and spatially inhomogeneous deformations, resulting from the rise and then fall in temperature that occurs in the fusion zone, the HAZ and the base material. These stresses exist after cooling down without any application of external load or momentum [11] and are of the constraint or reaction types [12]. The appearance of residual stresses during welding is inevitable and their presence affects the performance of the welded joint. Stress corrosion cracking [13], fatigue life [14], and toughness [15] in different regions of the joint are all influenced by the development of residual stress, as well as the microstructure. Residual stresses can be divided into three categories, organized with regards to their spatial extent (Fig. 1.5) [16]. Type I is macroscopic in nature and occurs over several grain lengths. Type II is a structural micro stress and covers the distance inside a grain and occurs between different phases or between inclusions and matrix. Type III is the residual stress developed inside a grain, over several

Fig. 1.5 Three kinds of residual stresses developed in metals

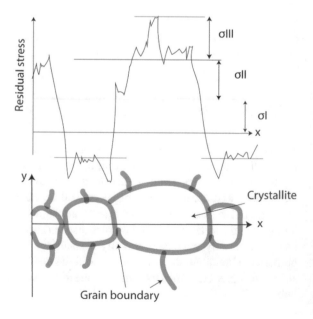

atomic distances. Simulation procedures described in this book are concerned with the first type of residual stress, which are macroscopic in size.

These three types of residual stresses can be obtained according to the following equations [17]:

$$\sigma^I = \frac{1}{V_{macro}} \int \sigma(x)dV \tag{1.3}$$

$$\sigma^{II} = \frac{1}{V_{crystallite}} \int \left(\sigma(x) - \sigma^I\right)dV \tag{1.4}$$

$$\sigma^{III}(x) = \sigma(x) - \sigma^I - \sigma^{II} \tag{1.5}$$

where V_{macro} and $V_{crystallite}$ are the volume of the macro region concerned and the volume of the crystallite, respectively. The phase residual stress (σ^α) refers to the sum of residual stresses of types 1 and 2 in an individual phase and is obtained from:

$$\sigma^\alpha = \left(\sigma^I + \sigma^{II}\right)^\alpha \tag{1.6}$$

The micro residual stresses in all phases (n number of phases) should be in equilibrium and therefore:

$$\sum_{\alpha=1}^{n} p^\alpha \sigma^{II} = 0 \tag{1.7}$$

where p^α is the volume fraction of each phase. The weighted average phase residual stresses is the macro residual stress and is obtained from:

$$\sum_{\alpha=1}^{n} p^\alpha \sigma^\alpha = \sigma^I \tag{1.8}$$

In order to measure the residual stresses, three experimental approaches can be used; destructive, semi-destructive and non-destructive [18]. The examples are block sectioning, hole drilling and neutron diffraction, respectively [19]. Various methods exist to measure the residual stress during welding. The neutron diffraction method is non-destructive and uses the interplanar lattice spacing as a strain gage [20]. Residual stress measurement by neutron diffraction is based on Bragg's law. The lattice spacing (d) is obtained according to:

$$2d \sin \theta = \lambda \tag{1.9}$$

λ is the wavelength of the neutron and 2θ is the diffraction angle. The lattice spacing in a stress-free material is d_0 and the corresponding diffraction angle is $2\theta_0$. Then the strain (ε) is calculated from:

$$\varepsilon = \frac{d - d_0}{d_0} = -\cot\theta.(\theta - \theta_0) \tag{1.10}$$

The residual elastic strain in orthogonal direction is obtained according to the Hook law:

$$\sigma_i = \frac{E}{1+v}\varepsilon_i + \frac{vE}{(1+v)(1-2v)}\left(\varepsilon_x + \varepsilon_y + \varepsilon_z\right) \tag{1.11}$$

where the indexes i = x, y, and z are the spatial directions and E and v are the Young's modulus and Posson's ratio, respectively. The non-destructive tests only measure the residual stress at the surface and the measurement of residual stress inside the material is not accurately measured by non-destructive methods.

The other method, which is destructive in nature, is the sectioning method introduced by Kalakoutsky [21] in which a small piece is cut away from the workpiece. Freed from the workpiece, this part will deform under the influence of its residual stresses. The strains released after cutting can be measured by strain gauges attached to the specimen before cutting. Figure 1.6 shows schematically the variation of dimensions after sectioning by which the strain in x and y directions can be obtained. The residual stresses are correlated to the measured strains by the following equations:

$$\sigma_x = -\frac{E}{1-v^2}\left(\varepsilon_x + v\varepsilon_y\right) \tag{1.12}$$

$$\sigma_y = -\frac{E}{1-v^2}\left(\varepsilon_y + v\varepsilon_x\right) \tag{1.13}$$

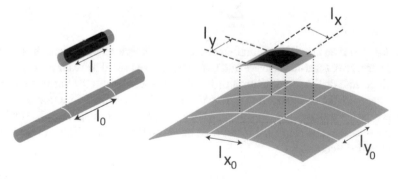

Fig. 1.6 Principle of sectioning method for measurement residual stress in (left) one dimension and (right) two dimensions

An alternative method uses simple analytical formulae, as suggested by Umemoto [22–24] or by Leoni [25] which are simple and quick to use, but are not as precise and are restricted to limited geometries and conditions. Numerical approaches are currently extensively used to calculate the residual stresses and its distribution over the desired area. Experimental methods can only measure the surface residual stresses and are not as cost effective if one desires to measure the distribution of residual stress. A numerical model can provide the residual stress distribution throughout the thickness, something which is very hard to measure with experiments. Furthermore, the effect of the welding parameters on the residual stress can be efficiently determined by simulation, while a comparable experimental approach would be very costly and long. This simulation process only requires a calibration of the thermal analysis for each parameter [26]. Simulation thus greatly reduces costs, since only a limited number of experiments is required to validate the model. In addition, it also provides a detailed distribution of the residual stresses at each point in the most critical (and experimentally inaccessible) zones. However, as welding is a transient non-linear problem, a long computation time may be needed to gather all the necessary data-points for all process stages [27]. Several approaches have been developed to reduce the computation time for the finite element (FE) analysis to predict the residual stress without compromising the accuracy of the model. Chapter 4 discusses various approaches that are used to perform the mechanical analysis for calculating residual stress and distortion precisely and in a shorter time.

Distortion during welding is often accompanied by residual stress. Many complex mechanical structures are composed of several components joined by welding, where every component has its own stiffness which determines the global stiffness of the structure. The higher the stiffness, the most severe is the restriction on the allowable volumetric or dimension change which reduces the distortion and increases the residual stress. Usually, factors that promote residual stress reduce the distortion and vice versa [28], as can be observed in Fig. 1.7.

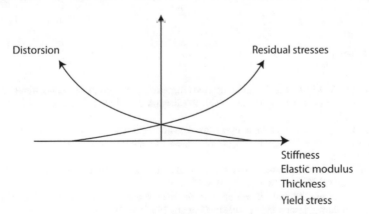

Fig. 1.7 Factors that promote residual stress and distortion

Distortions which result from welding interfere with the performance of the components, leading to important changes of the local dimensions. Thus, it is necessary to minimize these distortions by enacting adequate provisions. Changing the welding sequence and/or constrains are the most important methods to control the distortion. Experimentally, this needs many measurements to be carried out, which is again a very slow, inefficient and costly process. The use of simulation of these processes reduces both time and cost, as it only requires a limited set of experiments for calibration purposes. Furthermore, a calibrated model can be used to optimize the process parameters, allowing to obtain the desired objective.

In the automotive industry, where a vehicle body is assembled by joining several non-rigid sheets by spot welding, the final geometry is strongly influenced by the welding sequences [29]. Assessing the performance of all weld sequence strategies via experiment in order to obtain the optimum functional geometry is practically impossible and even the simulation of all these sequences requires a deep and comprehensive analysis which consumes long periods of time. For example, for n spotwelds, there will be n! available welding sequences. Therefore, optimization methods are indispensable to reduce time and cost. In Chap. 5, optimization methods used to reduce the calculation time and the number of runs are discussed in greater detail. In finite element simulation, diverse methods and simplifications have been created and can be applied to solve this issue and decrease the time of calculation. However, one must be aware that while these methods can be highly advantageous for industrial applications, many of the simplifications put in place greatly reduce the scientific potential of the models and only limited conclusions may be derived from these simplified simulations. In optimization problems, most of the time, understanding the real physical phenomena which leads to residual stress development is not important. However, in welding simulation processes it is important to construct a reliable model that can predict residual stress and/or distortion with acceptable accuracy. In this way, optimization algorithms can be utilized to design processes which lead to minimal residual stress or distortions during and after welding.

References

1. Huang, R.-S., Liu, L.-M., Song, G.: Infrared temperature measurement and interference analysis of magnesium alloys in hybrid laser-TIG welding process. Mater. Sci. Eng. A **447**(1–2), 239–243 (2007)
2. Naksuk, N., et al.: Real-time temperature measurement using infrared thermography camera and effects on tensile strength and microhardness of hot wire plasma arc welding. Metals **10**(8), 1046 (2020)
3. Chen, S., et al.: Temperature measurement and control of bobbin tool friction stir welding. Int. J. Adv. Manuf. Technol. **86**(1), 337–346 (2016)
4. Wang, L., et al.: Precipitates evolution in the heat affected zone of Mg-Gd-Y-Zr alloy in T6 condition during laser welding. Mater. Charact. **154**, 386–394 (2019)
5. Jeong, S., et al.: Influence of κ-carbide precipitation on the microstructure and mechanical properties in the weld heat-affected zone in various FeMnAlC alloys. Mater. Sci. Eng. A **726**, 223–230 (2018)

6. Li, X., et al.: Structure and crystallography of martensite–austenite constituent in the inter-critically reheated coarse-grained heat affected zone of a high strength pipeline steel. Mater. Charact. **138**, 107–112 (2018)
7. Nunes, A.: An extended Rosenthal weld model. Weld. J. **62**(6), 165s–170s (1983)
8. Promoppatum, P., et al.: A comprehensive comparison of the analytical and numerical predic-tion of the thermal history and solidification microstructure of Inconel 718 products made by laser powder-bed fusion. Engineering **3**(5), 685–694 (2017)
9. Wenji, L., et al.: A kind of analytical model of arc welding temperature distribution under varying material properties. Int. J. Adv. Manuf. Technol. **81**(5), 1109–1116 (2015)
10. Jeong, S., Cho, H.: An analytical solution to predict the transient temperature distribution in fillet arc welds. Weld. J.-Including Weld. Res. Suppl. **76**(6), 223s (1997)
11. Wohlfahrt, H., Macherauch, E.: Die Ursachen des Schweißeigenspannungszustandes. Mater. Test. **19**(8), 272–280 (1977)
12. Kannengießer, T.: Untersuchungen zur Entstehung schweißbedingter Spannungen und Verfor-mungen bei variablen Einspannbedingungen im Bauteilschweißversuch. Shaker (2000)
13. Khalifeh, A.: Stress corrosion cracking damages. In: Failure Analysis. IntechOpen (2019)
14. Li, L., et al.: Experimental and numerical investigation of effects of residual stress and its release on fatigue strength of typical FPSO-unit welded joint. Ocean Eng. **196**, 106858 (2020)
15. An, G., et al.: Evaluation of brittle fracture toughness by influence of residual stress. In: The 28th International Ocean and Polar Engineering Conference. OnePetro (2018)
16. Khanna, S.K., He, C., Agrawal, H.N.: Residual stress measurement in spot welds and the effect of fatigue loading on redistribution of stresses using high sensitivity Moiré interferometry. J. Eng. Mater. Technol. **123**(1), 132–138 (2001)
17. Spieß, L., et al.: Moderne röntgenbeugung: röntgendiffraktometrie für materialwissenschaftler, physiker und chemiker. Springer (2019)
18. Maekawa, A., et al.: Fast three-dimensional multipass welding simulation using an iterative substructure method. J. Mater. Process. Technol. **215**, 30–41 (2015)
19. Withers, P., et al.: Recent advances in residual stress measurement. Int. J. Press. Vessels Pip. **85**(3), 118–127 (2008)
20. Wan, Y., et al.: Weld residual stresses in a thick plate considering back chipping: neutron diffraction, contour method and finite element simulation study. Mater. Sci. Eng. A **699**, 62–70 (2017)
21. Ren, S., et al.: Finite element analysis of residual stress in 2.25 Cr−1Mo steel pipe during welding and heat treatment process. J. Manuf. Process. **47**, 110–118 (2019)
22. Umemoto, T., Furuya, S.: A simplified approach to assess weld residual stress distribution through pipe wall. Nucl. Eng. Des. **111**(1), 159–171 (1989)
23. Jonsson, M., Josefson, B.L.: Experimentally determined transient and residual stresses in a butt-welded pipe. J. Strain Anal. Eng. Des. **23**(1), 25–31 (1988)
24. Umemoto, T.: Tig torch re-heating for weld residual stress improvement in girth weld of thin pipe (1985)
25. Leoni, F., et al.: Rapid calculation of residual stresses in dissimilar S355–AA6082 butt welds. Materials **14**(21), 6644 (2021)
26. Hensel, J., Nitschke-Pagel, T., Dilger, K.: On the effects of austenite phase transformation on welding residual stresses in non-load carrying longitudinal welds. Weld. World **59**(2), 179–190 (2015)
27. Maekawa, A., et al.: Evaluation of residual stress distribution in austenitic stainless steel pipe butt-welded joint. Q. J. Jpn. Weld. Soc. **27**(2), 240s–244s (2009)
28. Nitschke-Pagel, T., Wohlfahrt, H.: Eigenspannungen und Schwingfestigkeit von Schweissverbindungen-eine Bewertung des Kenntnisstandes. HTM. Härterei-technische Mitteilungen **56**(5), 304–313 (2001)
29. Sadeghi Tabar, R., et al.: A novel rule-based method for individualized spot welding sequence optimization with respect to geometrical quality. J. Manuf. Sci. Eng. **141**(11), 111013 (2019)

Chapter 2
Heat Sources and Thermal Analysis

Abstract A key element of a successful welding simulation is the right selection of the model used as the heat source. The characteristics of the heat source, such as geometry, dimensions and heat value, must be selected according to the welding process and the weld geometries. This chapter describes various models that have been used so far to represent different welding processes and geometries. Arc welding, resistance welding, beam welding and friction stir welding (FSW) processes as well as common welding, narrow gap welding and key-hole welding are discussed. The validation methods of the models used for the heat source are also explained. In this way, a reliable thermal modelling of the welding processes can be achieved by considering the thermal boundary conditions and various mechanisms of heat dissipation such as conduction, convection, and radiation. The output of the thermal modelling is then suitable to be used in mechanical modelling in a subsequent stage.

2.1 Analytical Models

Analytical solutions proposed for calculation of temperature distribution for moving heat sources are all derived from a quasi-steady state condition. The most well-known analytical formula used to calculate the temperature around a moving heat source is the Rosenthal equation which calculates the temperature in the steady state system with respect to the moving coordinate system which is centralized with a point heat source. In this approach, a point-moving source is used for simple conduction welding and a line-moving source is used for welding processes with a deep penetration keyhole. Several modifications have been developed to improve the accuracy of these analytical solutions. For example, for keyhole welding, a line and several points with different heat levels have been defined along it in order to correctly predict the shape of the keyhole and the temperature distribution (Fig. 2.1). An analytical model for such a situation is proposed in [1]:

© The Author(s), under exclusive license to Springer Nature Switzerland AG 2022
R. Beygi et al., *Computational Concepts in Simulation of Welding Processes*,
SpringerBriefs in Computational Mechanics,
https://doi.org/10.1007/978-3-030-97910-2_2

Fig. 2.1 Laser welding
keyhole reference scheme

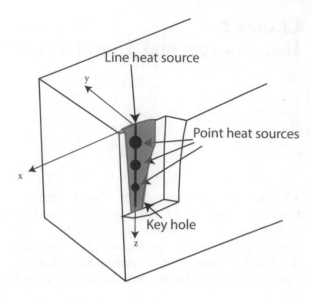

$$T(x, y, z) = T_0 + \frac{Q_L}{2\pi k} e^{-\frac{v}{2\alpha}x} K_0\left(\frac{vr_L}{2\alpha}\right) + \sum_{i=1}^{n} \frac{Q_{P,i}}{c_i \pi k r_{P,i}} e^{-\frac{v}{2\alpha}\left(r_{P,i}+x\right)} \qquad (2.1)$$

where $T(x, y, z)$ is the spatial distribution of temperature, T_0 is the initial temperature, Q_L is the line source unit strength per unit length, K_0 is the zero-order modified
Bessel function, r_L is the radial distance from the line heat source, v is the heat source
speed, α is diffusivity, $r_{P,i}$ is the radial distance from the ith point heat source, $Q_{P,i}$
is the heat strength of the ith heat source, k is the conductivity, and c_i determines the
strength of each point heat source. (x, y, z) is the coordinate system fixed to the heat
source. If one desires to measure the temperature in a fixed point on the workpiece,
it is necessary to transform the x to x':

$$x \rightarrow x' = x - vt \qquad (2.2)$$

where t is time.

The parameters used in the analytical formula such as the line of the heat source,
the number of point heat sources, the position of the heat sources and the coefficients should be calibrated. For this purpose, the extent and geometry of the weld
pool and the HAZ zone, which were obtained through an analytical solution, were
matched with the simulation results [1]. The most important advantage of the analytical formulations is that the influence of parameters on the heat distribution can
be quickly recognized by varying the parameter value in the formula [2]. A higher
number of variables make the analytical model more flexible. Although modifications can improve the reliability of analytical solutions for predicting temperature
distribution, they still lack the potential to be used for a wide variety of geometries,

since these equations hold only for semi-infinite or infinite plates. For example, it would not be as easy to determine the analytical solutions for finite plates or for more complex geometries like T-joints. In contrast, numerical simulation offers the possibility of defining the heat source with high flexibility in terms of its shape and number of parameters. Instead of solving complex equations for measuring the temperature distribution analytically, the parameters of the heat source are calibrated in this way and the temperature distribution is numerically solved by computer.

2.2 Heat Source Model for Numerical Simulation

The aim of the heat source simulation is not to simulate the welding process itself, but to effectively recreate the same thermal state as the real state. Therefore, the models used to simulate the heat source do not necessarily explain the real physics of the heat source. In direct arc welding processes, heat is generated by discharging the electrons at the anode and cathode as well as in the plasma column. In indirect arc welding processes, the heat is transferred through a plasma jet with a high velocity. The properties of the heat source must be adjusted in order to obtain the actual thermal state in the workpiece. The parameters that are used for the welding heat source simulation are the heat quantity (Q in J) and the heat flow (q in J/S) in the cases of spot welding and continuous welding, respectively. These quantities should be multiplied by a constant known as thermal efficiency (η). This thermal efficiency considers the heat losses due to convention and radiation of heat to the environment, the loss due to splash, and the loss due to the heating of the non-consumable electrode. In other words, η is the ratio of heat introduced in the workpiece to the thermal equivalent of the electric power of the arc. For example, the amount of heat in arc welding processes is obtained from:

$$q = \eta \frac{U.I}{V} \qquad (2.3)$$

where U is voltage, I is current, and V is the speed of the weld. The value of η differs for various kinds of arc welding processes due to different heat efficiency of them. For laser welding the generated heat is obtained from:

$$q = \eta \frac{P}{V} \qquad (2.4)$$

where P is the power of the laser welding. For a thermal analysis, it is essential to know the heat distribution in the workpiece in order to be able to implement it in the numerical simulation. For this purpose, several models were defined as representative of the heat sources for different types of welding processes that are to be used as heat sources in the numerical simulation. In the following subsections some of the most frequently used models are described in greater detail.

2.2.1   Arc Welding Processes

So far, several models have been proposed to simulate the heat sources in various welding processes. The heat flow density which is defined either per area (J/mm^2 s) or per volume (J/mm^3 s) has various distribution curves depending on the process. The coordinate system which is used for the heat source is critical. Furthermore, temperature distribution in the steady state and the transient state are different. In the stationary state, the coordinate system is attached to the heat source and the observer, who is fixed on the moving coordinate system, sees a constant temperature distribution that does not change over time. In fact, the temperature only depends on the location in relation to the moving coordinate system. In the transition state, like the beginning of the welding or spot welding, the temperature distribution around the heat source is time-dependent. Fourier's law states that the heat flow is related to the transient heat temperature in the domain. The temperature which is spatially and temporally distributed, T(x, y, z, t), can be calculated by solving the following equation [3]:

$$\rho c_p \frac{\partial T}{\partial t} - k \nabla^2 T = q \tag{2.5}$$

where q represents the heat source or heat sink in the domain, k is thermal conductivity, ρ is density, and c_p is the specific heat capacity. The most known model for q is a double-ellipsoidal heat source developed by Goldak [4]. A double-ellipsoidal model simulates the heat power density over a flat surface which is typical to arc welding processes which an arc impinges on a flat plate. In this model, the heat decays as the distance from the center of the heat source increases laterally. The 3D double-ellipsoidal heat source is shown in Fig. 2.2. In the 3D model, the extent of the heat distribution in the lateral direction decreases with increasing distance from the top. It consists of two ellipsoids, one leading and one trailing. The front heat source for a 3D heat source is defined as follows [5]:

$$q_f = \frac{6\sqrt{3} f_f Q}{abc_f \pi \sqrt{\pi}} e^{\left(-\frac{3x^2}{a^2}\right)} e^{\left(-\frac{3y^2}{b^2}\right)} e^{\left(-\frac{3z^2}{c_f^2}\right)} \tag{2.6}$$

and the rear heat source is given by:

$$q_f = \frac{6\sqrt{3} f_r Q}{abc_r \pi \sqrt{\pi}} e^{\left(-\frac{3x^2}{a^2}\right)} e^{\left(-\frac{3y^2}{b^2}\right)} e^{\left(-\frac{3z^2}{c_r^2}\right)} \tag{2.7}$$

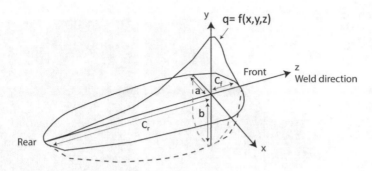

Fig. 2.2   Double ellipsoidal heat source proposed by Goldak

In which x, y, and z are the local coordinate system as shown in Fig. 2.2. The parameters (a, b, c_f, c_r) used in Eqs. 2.6 and 2.7 are defined in Fig. 2.2. f_f and f_r represent the fraction of the heat deposited in front and rear of the ellipsoid, respectively.

These parameters should be determined depending on the welding process, welding parameters, and geometry of the joint. In this regard, in every specific case, these parameters should be calibrated by adopting a proper approach. The parameters of the heat source can be fitted by three different approaches [6, 7]. First, by measuring the temperature at certain points during welding and iteratively adjusting the parameters of the heat source in order to match the temperature in the FEM analysis with the measured one. Second, by measuring the residual stress at selected points and iteratively adjusting the parameters in order to match the residual stress in the FEM simulation with experiments. This approach requires a mechanical model to be created. Third, through macrograph analysis of the welds to match the shape and size of both the melt zone and the heat affected zone with those of the FEA analysis. The width and depth of the melt zone are the most common features on the weld macrograph used to calibrate the heat source parameters [8]. An example of the matching of the weld seam shape obtained from the simulation with that obtained from the experiment is shown in Fig. 2.3. In thermal analysis, the boundary of the melting zone is defined as those areas whose local temperatures exceed the melting temperature. In this figure, a good match can be observed with respect to the boundary of the fusion zone. During the thermal analysis, the geometry of the joint is predefined, which also takes the added filler material into account. This simplification in the simulation means that the real shape of the weld pool and the resulting defects such as undercuts cannot be predicted. The geometric discontinuities cause a large concentration of stress, which influences the stress distribution in the real weld seam and must be determined by experimental methods. However, the boundary of the fusion zone can be accurately obtained by thermal simulation.

The prediction of the weld seam shape must take into account the liquid flow in the weld pool. The forces that act upon this flow during welding are Marangoni force, capillary pressure, buoyance, electromagnetic force and gravity. The electromagnetic force is exerted by the arc current on the molten pool. This force can be downward on

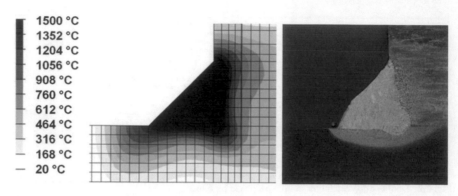

Fig. 2.3 Fitting of the weld seam shape obtained from the simulation with that one obtained from the experiment [9]

the surface of the pool and inward in the radial direction. The force of the arc acting on the weld pool such as vertical pressure pushes the melt downward. As the melt is pushed downward its radius decreases and the resistant pressure resulted from the surface tension of the melt increases. The other force is the shear force. The resistance shear force of the weld pool is in parallel to the surface and is low. The shear stress of the plasma, forces the weld pool to the back [10]. The Buoyancy force is due to the variation of density of the liquid resulting from temperature difference. The surface tension is the other force acting in the weld pool. Marangoni force occurs along an interface between two liquids with different surface tension. With this mechanism, a liquid with a high surface energy pulls the surrounding liquid with a lower surface energy. In other words, in the presence of a surface energy gradient, a flow of liquid occurs whose speed (u) is equal to:

$$u = \frac{\Delta\gamma}{\mu} \tag{2.8}$$

where μ is the viscosity, and $\Delta\gamma$ is the difference of surface tension between the liquids. The surface tension of the liquid metal depends on the chemical composition and temperature. In the weld pool, there is a gradient of temperature which influences the solubility of elements and therefore a gradient of surface tension exists that induces a flow of molten material [11] as shown in Fig. 2.4. The shape of the weld pool depends on the gradient of surface tension.

Modelling the flow in the weld pool requires the introduction of a computational fluid flow model in the simulation, which is mainly used to predict the shape of the weld pool in keyhole welding processes [12]. In fusion welding processes, there is a liquid weld pool, the dynamic flow of which is neglected by the numerical thermal simulations. Thermal analyses only calculate the temperature distribution, since taking the flow into account greatly increases the computing time. The flow in the welding pool affects both the shape of the weld pool and the heat dissipation. Heat

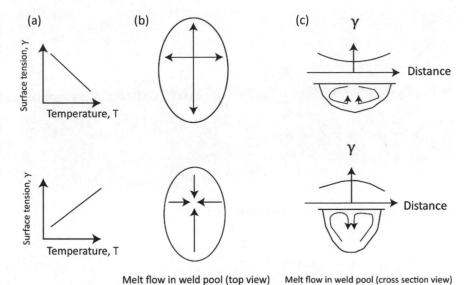

Fig. 2.4 Marangoni effect in the weld pool. **a** Variation of surface tension with respect to temperature. **b** Melt flow from top view. **c** Melt flow from cross section view and variation of surface tension in the weld pool

dissipation through the melt flow in the melt pool can be taken into account in the thermal analysis by assigning a higher thermal conductivity to the melt pool, above the melting temperature [13]. In thermal analysis it is also necessary to consider the heat loses due to convection and radiation as thermal boundary conditions. The heat loss due to convection (q_c) and radiation (q_r) are obtained from:

$$q_c = h_c(T - T_0) \tag{2.9}$$

$$q_r = \epsilon \, \sigma \left(T^4 - T_0^4\right) \tag{2.10}$$

where T_0 is the temperature of environment, h_c is the coefficient of the heat transfer, ϵ is the emissivity factor, and σ is the Stefan–Boltzmann constant.

In addition to the Goldak heat source, various other heat sources have been introduced and used in thermal simulation of welding. One common option for heat source simulation is a gaussian heat source with a bi-elliptical shape [14]. The heat source can be assumed to be 2D acting on the upper plane of the sheet (Fig. 2.5). This model is accurate enough to simulate the welding of thin sheets both by arc welding or laser welding [15]. In 2D models, the heat distribution along the thickness cannot be well estimated. The gaussian heat distribution is depicted in Fig. 2.5. The 2D heat distribution for laser welding obtained from the gaussian formula is expressed as [15]:

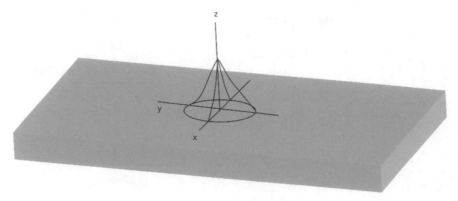

Fig. 2.5 Gaussian heat source with a bi-elliptical shape

$$q_{2D}(x, y) = \frac{2\eta P_{laser}}{\pi R^2} \left[e^{-\frac{2r^2}{R^2}} \right] \tag{2.11}$$

where P_{laser} is the power of laser, η is the heat efficiency, R is the radius of the heat source which is defined as the area in which 95% of heat is transferred to the work, and r is the distance to the center of the heat source. A similar heat source can be defined for arc welding processes.

The use of 3D heat sources often yields more accurate results and several strategies can be utilized to create them. One of these is the use of composite heat sources, where a volume heat flux is used along with a surface heat flux. Another heat source model is the three-dimensional conical heat flux shown in Fig. 2.6. This is a volumetric heat source which takes into account the heat distribution along the thickness. At radial distances perpendicular to the axis of z the heat distribution $(q(r, z))$ is obtained

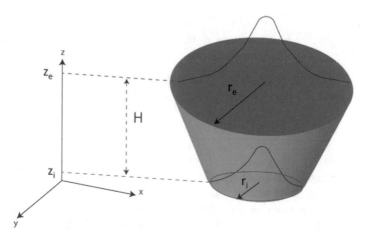

Fig. 2.6 Three-dimensional conical heat flux

from:

$$q(r, z) = \frac{9\eta U I r^3}{\pi \left(e^3 - 1\right)(Z_e - Z_i)\left(r_e^2 + r_e r_i + r_i^2\right)} e^{-\left(\frac{3r^2}{r_0^2}\right)} \qquad (2.12)$$

where r_0 is the distribution profile, r is the radial coordinate, z_e and z_i are the z coordinates of the top and bottom surfaces, respectively, and r_e and r_i are radius at the top and bottom surfaces, respectively. The distribution profile r_0 decreases linearly from the top to the bottom of the conical region (r_e at top and r_i at bottom).

In plasma arc welding (PAW), a plasma jet with a very high velocity (in the order of 300–200 m/s) forms a keyhole in the work piece. The methods used to simulate the keyhole are volume of fluid (VOF) and level set (LS). The gaussian surface flux does not consider the key hole effect [16]. The prediction of the weld pool, which is produced by keyhole welding or welding within deep narrow grooves, is a complicated procedure and using the double-ellipsoidal heat power density distribution will not be accurate. Narrow groove welding is used for thick sections of materials and is different from typical arc welding processes in which an arc impinges on the surface of the workpiece. In flat welding, a part of the heat is dissipated through convection and the other part is dissipated through radiation from the surface of the work. In the base of a narrow groove, a part of the heat of the arc is transferred to the walls of the groove through radiation and forced convection, as shown in Fig. 2.7.

Usual heat source models do not take into account the heat transfer to the walls of narrow groove via convention and radiation. For more accurate results, a double-ellipsoidal-conical heat power distribution can be used, as shown in Fig. 2.8 [3]. The double-ellipsoidal heat source accounts for the base of the groove, like what was observed for a flat surface. The double-conical heat source accounts for the wall of the groove into which the heat of the arc is transferred by radiation and forced convection. 'y_i' is the position of the heat source. When an arc is moving on the flat surface and no groove is present, this model changes to a simple double ellipsoidal one [3]. Therefore, this model can be used to simulate the heat source on the flat surface as well as a full penetration or a partial penetration in the case of keyhole or a narrow gap penetration. In narrow gap penetration, as several passes are used, the heat efficiency and the dimensions should be adopted in each pass [3]. The heat

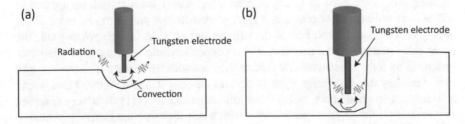

Fig. 2.7 Schematic of heat dissipation of arc on a **a** flat surface and **b** within a narrow groove

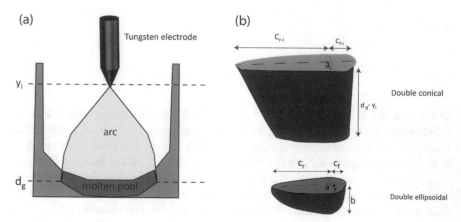

Fig. 2.8 a Schematic of the deep narrow groove weld. **b** The double-ellipsoidal-conical heat source model used for the deep narrow groove weld [3]

efficiency in the narrow groove welds should be changed and fitted in each pass, since as the number of passes increases, d_g decreases and the stifling effect for radiation and convection decreases, leading to a higher heat loss [3].

The simulation of the heat source in TIG welding is strongly dependent on the shape of the heat source. A study on this subject employed three different heat source shapes: Two-dimensional Gaussian, three-dimensional conical and three-dimensional Goldak [17]. Various combinations of materials were welded by TIG and the resultant weld bead profile and time–temperature histories were compared with simulated ones using three different heat sources. The temperature profiles obtained by each heat source are shown in Fig. 2.9. Generally, three-dimensional heat sources result in more accurate prediction of the welding pool boundary, especially at the root of the weld.

2.2.2 Laser Welding

The mechanism of arc welding processes is different from those of beam-based welding processes, such as laser beam welding. Laser welding can be applied in different two modes: low power and high power. In low power mode, most of the energy of the beam is absorbed on the surface and melts it. In high power mode the beam fully penetrates the material, and a deeper keyhole is formed. In a keyhole produced by a laser beam several factors interact with each other. The metal vapor inside the keyhole can dampen the beam and defocus it. On the other hand, it can also cause a thermal concentration due to plasma radiation [1]. It is very complex to simulate all the physical phenomena which occur during the keyhole formation by laser welding. If the intention is only to obtain the temperature distribution, the keyhole can be assimilated with a virtual heat source. Ostuni et al. [15] considered

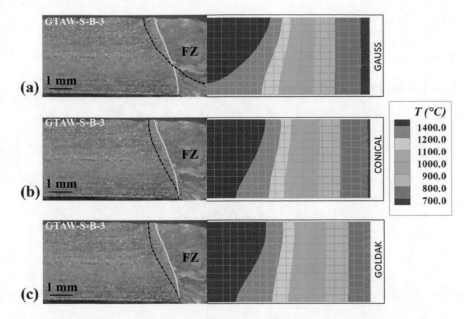

Fig. 2.9 (Left) Macrostructure of the weld showing the fusion boundary obtained by experiment (Yellow line) and numerical analysis (Black dotted line). (Right) Temperature profiles obtained by simulation using **a** 2D-Gaussian, **b** 3D-Conical, and **c** 3D-Goldak heat sources [17]. Reprinted with permission from Elsevier

a cylindrical heat source in the thickness with a radius equal to the radius of the key-hole measured by experiment. In this way, a Gaussian heat source with radius R_{FZ} along the thickness was considered. This heat source is in addition to the 2D-Gaussian heat source previously applied on the surface (Fig. 2.10). A fraction of heat $(1 - \varphi)$ is provided from the surface heat source and a fraction (φ) is provided from the volumetric heat source. The total heat (hf_{3D}) is related to the surface heat source (hf_{sur}) and the volumetric heat source (hf_{vol}) according to the following equation:

Fig. 2.10 Cylindrical plus a Gaussian surface heat source used to simulate the keyhole in laser welding

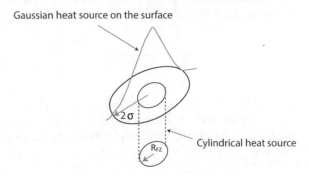

Fig. 2.11 Selection of a
coordinate system for a laser
beam heat source used in a
fillet weld [18]

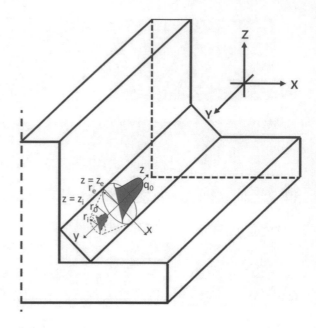

$$hf_{3D} = (1 - \varphi)[hf_{sur}(x, y)] + \varphi[hf_{vol}(z)] \tag{2.13}$$

where z is the thickness direction. For laser welding of a fillet weld, a cone heat
source was used to represent the key-hole effect whose distribution, according to
Fig. 2.11, is obtained from [18]:

$$q_r(r, z) = q_0 e^{-3\left(\frac{r}{r_0}\right)^2} \tag{2.14}$$

where r is the current radius of the interior point within the cone, r_0 is the radius
of the heat source at a specific z, and q_0 is the maximum volumetric power density.
The axis of the cone is perpendicular to the weld surface. The maximum volumetric
power density is obtained from:

$$q_0 = \frac{9\eta Qe^3}{\pi\left(e^3 - 1\right)(z_e - z_i)\left(r_e^2 + r_e r_i + r_i^2\right)} \tag{2.15}$$

where Q is volumetric heat flux. The dimensions are shown in Fig. 2.11.

For a fillet weld a local coordinate system is needed which is obtained by
multiplying the global coordinate system (X, Y, Z) by transformation matrix as
follows:

$$\begin{bmatrix} x \\ y \\ z \end{bmatrix} = \begin{bmatrix} \cos 45 & \sin 45 & 0 \\ -\sin 45 & \cos 45 & 0 \\ 0 & 0 & 1 \end{bmatrix} \begin{bmatrix} X - X_0 \\ Y - Y_0 \\ Z - Z_0 \end{bmatrix} \tag{2.16}$$

where X_0, Y_0 and Z_0 are the coordinate values of a point in the global coordinate system (X, Y, Z), as shown in Fig. 2.11.

The double-ellipsoidal conical heat source can also be used to model the heat power distribution in laser beam welding. According to the model shown in Fig. 2.8, even in the case of a full penetration, the double ellipsoidal-conical yields the better results in prediction of temperature–time history in the back of the plate [3]. The ellipsoidal part in the bottom yields better results when the penetration is partial. In the case of full penetration and keyhole, the double-ellipsoidal-conical model tends to be double-conical. In the case of electron beam welding in which the beam penetrates and exits from the bottom of the sheet, this ellipsoidal part is not useful and the heat source would be more like a double conical heat source [3]. A three-dimensional Gaussian conical heat source was used to simulate the heat source in laser welding [19]. Another model used for laser beam welding is an hour-glass like heat source in which the heat distribution is Gaussian and is obtained separately for the upper and lower parts from [20]:

$$Q_v(x, y, z) = \begin{cases} \frac{9Q_0}{\pi(1-e^{-3})} \cdot \frac{1}{(z_e - z_0)(r_e^2 + r_e r_0 + r_0^2)} \cdot \exp\left(\frac{-3r^2}{r_{c1}^2}\right) & z_0 < z < z_e \\ \frac{9Q_0}{\pi(1-e^{-3})} \cdot \frac{1}{(z_e - z_0)(r_e^2 + r_e r_0 + r_0^2)} \cdot \exp\left(\frac{-3r^2}{r_{c2}^2}\right) & z_i < z < z_0 \end{cases} \tag{2.17}$$

The shape of this heat source is shown in Fig. 2.12. The geometrical parameters of the model as well as the heat input should be calibrated by comparison with experiments and utilizing the optimization methods. The hourglass-shaped heat source has the unique potential to predict the wider bottom weld width created by full penetration laser welding [21].

Further modifications have been carried out to use double conical heat source for simulation of laser welding. A process where tailor welded blanks of two dissimilar

Fig. 2.12 Hourglass heat source used for laser beam welding [20]. Reprinted with permission from Elsevier

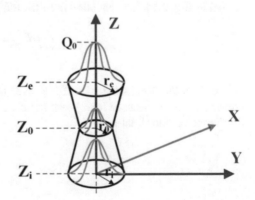

Fig. 2.13 The heat source
comprising of two half
hourglass heat sources with
truncated upper part

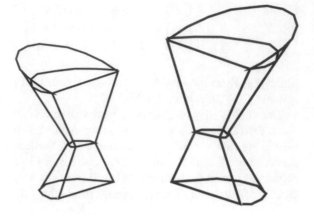

thicknesses are welded was simulated by using two half hourglass shape heat sources
with a Gaussian heat distribution [22] (Fig. 2.13). However, the geometrical shape
of the heat source is complex, and several parameters have to be optimized.

In electron beam welding, electrons penetrate the material further than the photons
in laser welding. In both a vapor capillary is formed that acts as a heat source. The
ratio of depth to width of the weld zone in electron beam welding is in general higher
than that achieved with laser welding. The complex interaction between material flow
in the weld pool, the heat distribution, and the shape of the weld pool necessitates
utilizing computational fluid dynamics. For general purposes of calculating residual
stress or temperature distribution, a heat source can be defined based on the weld
pool geometry and dimensions. The examples are a half hemisphere on top plus a
partial cone on the bottom [23] and 2D Gaussian plus 3D conical heat sources [24].

With low-power laser welding, the keyhole effect is no longer present, and the
heat is transferred to the workpiece via the top surface. Laser transmission welding
is a variant of laser welding process which is used to join thermoplastic components
in lap configuration. The top sheet is transparent to the beam and the lower one is
absorbent. When doing laser transmission welding, a two-dimensional heat source
can be realized at the interface where the beam is absorbed into the lower sheet as
shown in Fig. 2.14. The equation used for the heat source is defined as [25]:

$$q_0(x, y) = \frac{c.n.P}{\pi.r^2} e^{\left(-c\left(\left(\frac{x}{r}\right)^2 + \left(\frac{y}{r}\right)^2\right)\right)} \tag{2.18}$$

where P is the nominal power of the laser beam, n is the percentage of the heat
absorbed, r is the focused laser beam radius, c is a shape parameter, and x and y are
the Cartesian coordinates.

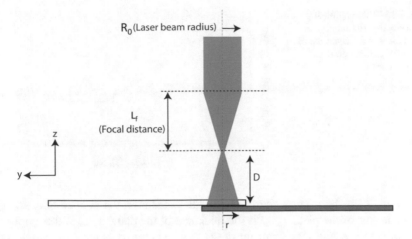

Fig. 2.14 Schematic of laser transmission welding heat source

2.2.3 Friction Stir Welding

The thermal analysis for solid-state welding processes follows the same rules as fusion welding processes if the material flow simulation is not a goal. Friction stir welding (FSW) is a solid-state welding process in which heat is produced by friction between a tool and the workpieces. This heat causes softening of the material and facilitates the plastic deformation of the material around the tool. The softened material, under high strain and high strain rate, flows around the tool and fills the cavity behind the tool, forming a joint. The approach to simulating this process depends on which outputs are important for a given design process. The main differences between this process and fusion welding processes are that no melting occurs during welding and no filler material is included in the welding process. Nonetheless, computational fluid dynamics can still be used to capture the material flow. A simple approach is to use a heat source instead of the tool, where the contact area of the pin and shoulder with the material is viewed as the heat source. In this way, the movement of the material during welding is not calculated and no data on strain and strain rate can be obtained and only the thermal analysis will be performed.

The main point of the heat source model used in FSW is that the maximum temperature of the process should be kept below the melting point of the material. One strategy is to include the coefficient of friction in the model and relate the coefficient of friction to the temperature [26]. The relationship between the friction coefficient and temperature is depicted in Fig. 2.15. In this way the heat is calculated as follows:

$$\dot{q}_{FSW} = \frac{2\pi}{3A_{node}t_{plate}N_{tool}}\omega\mu_k(T)pr_2^3 \qquad (2.19)$$

Fig. 2.15 The relationship
between the friction
coefficient of the interface of
the FSW tool-material and
temperature

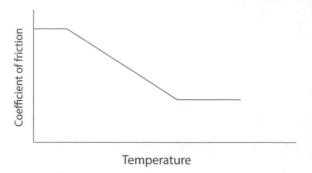

where $\mu_k(T)$ is the friction coefficient and dependent on the temperature, A_{node} is
the node area of the plate, t_{plate} is the thickness of the plate, N_{tool} is the number of
the nodes representing the diameter of the tool, p is the axial pressure and r_2 is the
radius of the shoulder.

A new solid state joining has been developed, named hybrid metal extrusion and
bonding (HYB) [27]. This process uses a filler material to fill the gap by extrusion
mechanism. In this regard it is similar to fusion welding but it is still performed in
solid state. In an attempt to simulate the process, a Goldak model was used for the
heat source likewise fusion welding process [28].

2.2.4 Resistance Spot Welding

Resistance welding is a non-stationary problem [29]. In addition to thermal effects,
the forces associated to the welding process can also influence the residual stresses.
The pressure applied by the electrode in resistance welding can induce compressive
residual stresses. An electrical-thermal-mechanical model is used to analyze the
resistance-based welding procedures. MSC. MARC is a nonlinear finite element
commercial software capable to do this. A mechanical model requires the thermal
field to determine the material properties and to obtain the thermal field, the heat
source needs to be known. For determination of the heat source, electrical models
are used to calculate the current distribution and the resultant heat. The coupling of
these three phenomena is depicted in Fig. 2.16 [30].

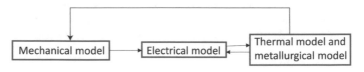

Fig. 2.16 Coupled electrical, thermal and mechanical model

The analysis can be decomposed into separate coupled analyses to simplify the simulation process. First, a thermal-electrical analysis is carried out to determine the temperature, which is then followed by a thermal-mechanical analysis [31]. In this way, the temperature field is obtained from thermal-electrical analysis and it is adopted as a body load in the thermal-mechanical analysis. The heat is produced due to contact electrical resistance ($\rho_{contact}$) and it is obtained from [30]:

$$\rho_{contact} = 3.\left(\frac{\sigma_{s.soft}}{\sigma_n}\right)\left(\frac{\rho_1 + \rho_2}{2} + \rho_{contaminants}\right) \qquad (2.20)$$

where $\sigma_{s.soft}$ is the yield stress of the softer material in contact, σ_n is the contact normal pressure, ρ_1 and ρ_2 are the resistance of the two materials in contact, and $\rho_{contaminants}$ is the resistance due to contamination such as oxides at the interface. The thermal and electrical contact can be modeled by introducing an artificial film at the interface [30]. The boundary condition for each model should be defined properly. The electrical potential at the bottom of the lower electrode is considered zero in thermal-electrical analysis [31]. The roughness of the contact area influences the heat generation and can be taken into account [32], as shown in Fig. 2.17.

The major part of the heat transfer in this process takes place by conduction to the water-cooled electrodes, ensuring that the heat transfer can be considered to occur only in one direction (electrode direction) [33]. The heat transfer equation for resistance welding is obtained from [34]:

$$D.c.\frac{\partial T}{\partial t} = \frac{\partial}{\partial x}\left[k.\frac{\partial T}{\partial x}\right] + \frac{\partial}{\partial y}\left[k.\frac{\partial T}{\partial y}\right] + \frac{\partial}{\partial z}\left[k.\frac{\partial T}{\partial z}\right] + \sigma\nabla\varphi.\nabla\varphi \qquad (2.21)$$

where D is the mass density, T is temperature, t is time, c is heat specific capacity, k is thermal conductivity, ϕ is the electrical potential, and σ is electrical conductivity. Various boundary conditions should be defined in both thermal and electrical models, such as the contact surface of electrode and sheet, contact surfaces of sheets (faying

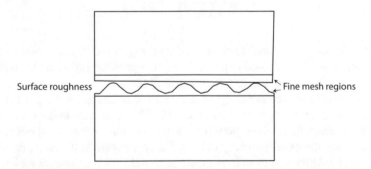

Fig. 2.17 Description of the model with considering the contact roughness in simulation of resistance spot welding

Fig. 2.18 Graphical
boundary conditions for
electrical, thermal, and
mechanical models

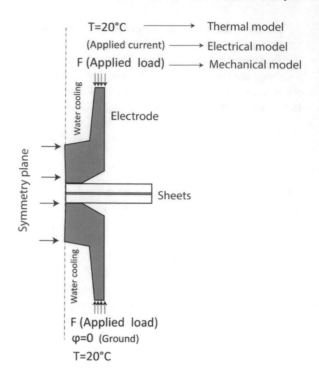

surfaces) and the surface exposed to ambient. The boundary conditions in electrical, thermal and mechanical models are presented in Fig. 2.18. In the thermal model a convective water cooling process (represented by h_w) exists within the electrodes and a convective air cooling process (represented by h_a) exists on the free surfaces of electrodes and plates. The heat is also conducted in plates, electrodes and the interfaces. The thermal conductivity at the interface (k_c) can be defined as follows [34]:

$$k_c - \frac{1}{3}\left(\frac{\sigma}{\sigma_e}\right)\left(\frac{k_1 + k_2}{2}\right) \tag{2.22}$$

where σ is the average normal stress, σ_e is the average yield stress, k_1 and k_2 are the thermal coefficients of contacting parts. The corresponding equations for boundary conditions in different models can be found in [34].

The verification of the accuracy of the FEM simulation can be performed by measuring the temperature of the electrode [35]. The other way, as mentioned, is by matching the shape of the weld pool (here known as the weld nugget) obtained via simulation with the experimentally obtained shape. Figure 2.19 shows the experimental and simulation results of a spot resistance weld made on two dissimilar steels (stainless steel and carbon steel) with the same thickness obtained by commercial software SORPAS [30]. SORPAS® [36] couples four models, considering electrical,

Fig. 2.19 The weld nugget shape obtained by experiment and simulation along with the temperature distribution in simulation [37]

thermal, metallurgical, and mechanical aspects [35]. As the simulation result shows, the weld nugget zone in stainless steel is larger, which is due to the lower heat conductivity of this steel.

2.2.5 Hybrid Welding Technologies

Hybrid welding processes such as hybrid laser-arc welding procedure are used to establish a balance between the heat delivery and the weld penetration. Common welding procedures such as arc welding processes used on thick sections require high heat input to achieve sufficient weld penetration. Intensive heat sources such as a laser can be used along with arc welding to provide a high penetration level. For modelling the heat source in this case, two heat sources corresponding to each process can be superimposed and used together. For hybrid laser-arc welding processes, a combined double ellipsoidal-conical heat source has been used in the literature [20]. The shape of this double heat source is shown in Fig. 2.20. For GMAW-Plasma hybrid welding a combined double-ellipsoidal shaped Goldak's plus two connected

Fig. 2.20 Combined double ellipsoidal-conical heat source for hybrid welding [20]. Reprinted with permission from Elsevier

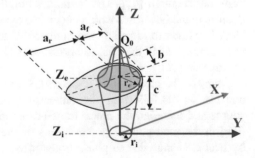

Pavelic's disc shaped models has been used as the heat source [38]. The first and second shapes represent GMAW and PAW, respectively.

2.3   Phase Transformation and Material Properties

The determination of the transient thermal field during welding is the most important prerequisite for predicting the phase transformation and the residual stress. Analytical models can quickly and roughly estimate the influence of some parameters on the cooling rate, but this approach cannot accurately provide the spatial distribution of the microstructure and residual stress [39]. The temperature field during the welding process is dependent on the way this heat is transferred and dissipated. As the welding processes are mostly concerned with metals, the most important mechanism of heat transfer in welding is conduction. Other mechanisms, such as convection and radiation, have a variable contribution to the heat transfer, depending mostly on the ambient temperature, geometry of the joint, dimension of the work piece, material of the work piece, and the surrounding materials in contact with the work piece such as backing plate or fixture.

The thermal analysis of the process via simulation offers the possibility to precisely record the thermal history at each location and thus the spatial distribution of the phases can be predicted with a high degree of accuracy. A correct thermal analysis by numerical simulation offers the possibility to plan ahead and control the formation of undesired phases in critical areas. In the case of high-strength low-alloy steels, the coarsening of the austenitic grains in HAZ, which is dependent on the peak temperature, reduces the toughness of the phases subsequently formed during cooling [40]. In addition, the cooling time between 800 and 500 °C ($t_{8/5}$) is very critical as it determines the microstructure of steel. In some cases, it is needed to control this cooling time to avoid the formation of detrimental microstructures (martensite and bainite) which may cause cracks. The simulation of thermal process provides this potential to measure this time in critical areas of the weld and thereby determine the percentage of each phase. Figure 2.21 shows the calculated $t_{8/5}$ and the calculated percentage of martensite.

For a thermal analysis, materials properties such as thermal conductivity, convection coefficient, density, specific heat, melting point, and initial component temperature are required [41]. The boundary conditions should be defined correctly to determine the heat loses which occur due to convection, radiation, and conduction. Heat loses due to radiation and convection are more important during welding of thin sections, as in the thick sections most of the heat is sunk into the material by conduction [42].

Many of the models used to determine the temperature field are applied without considering the following phenomena: the complex interaction between the heat source and material, the phase transformation of solid–liquid, and the flow of the fluid in the melt pool [13]. Variations of the temperature field which are caused by heat exchange due to phase transformations in solid state are negligible [13],

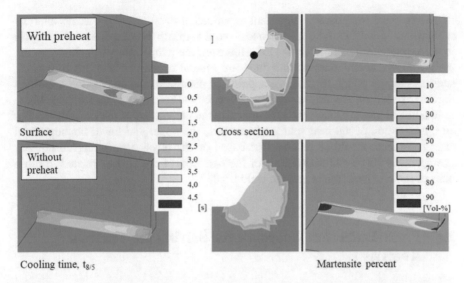

Cooling time, $t_{8/5}$ Martensite percent

Fig. 2.21 The effect of preheating on the cooling time ($t_{8/5}$) in various parts of the weld and the corresponding martensite percentage shown on the surface of the weld [13]. Reprinted with permission from the author

however, the solid–liquid phase transformation in the weld pool may influence the thermal analysis. This influence is due to both variation of materials properties and material flow inside the weld pool. In order to take these into account and at the same time avoid large increases in the complexity of the model, the following strategies can be implemented:

- Reduction of the specific heat capacity in melt temperature;
- Interpolation of the temperature above the melting interval;
- Variation of enthalpy above the temperature interval of transformation;
- Change of density and heat conduction coefficient.

In order to consider the convective motion in the weld pool, an artificial thermal conductivity can be considered for the liquid phase which is higher than the one found in the solid state [43].

2.4 Validation of Thermal Analysis

The de-coupled thermal and mechanical analysis assumes that these two analyses are weakly coupled, and that temperature distribution is not affected by stress and strain during welding. The obtained temperature history from thermal analysis is used as a thermal load input in mechanical analysis in which the stress and strain are obtained. Thermal strain induced by variation of temperature distribution causes deformation.

In a de-coupled analysis, it is important to validate the results of the thermal analysis with experimental data. One approach to do so is to match the temperature–time cycle in some predetermined locations [13]. However, the temperature–time histories are not usually perfectly matched and there are several reasons behind this mismatch, such as change of the measurement point or change in the convection condition during experiment. Another approach is to match the boundary of fusion zone which is considered as a reliable method to optimize the parameters of the heat source. For an optimization of the heat source parameters, the upper and lower boundaries of each parameter should be defined first based on the physical meaningfulness. For example, the shape and dimensions of the heat source are closely related to those ones of the real heat source and the weld pool.

2.5 Finite Difference Approach for Solving the Thermal Analysis

The basic equation for thermal analysis is:

$$[A]\{T\} + [C]\left\{\frac{\partial T}{\partial t}\right\} = \{F\} \tag{2.23}$$

[A] is the matrix of thermal conductivity, [C] is the matrix of heat capacity, {T} is the temperature field array, and {F} is the array of boundary condition and heat source. The approaches used for solving the temperature field equations based on numerical methods mostly use finite difference or finite element approaches. In the finite difference method, differential equations are solved by making an approximation in which derivatives are substituted with finite differences. In this way the spatial domain is discretized into several points and the value of temperature in every time interval is calculated in every point by considering the temperature in nearby points. For example, the discretization of a surface into points is depicted in Fig. 2.22. This figure represents a 2D heat transformation in which the temporal and spatial derivations are obtained from following equations:

$$\left.\frac{\partial T}{\partial t}\right]_{i,j} = \frac{T_{i,j}^{k+1} - T_{i,j}^{k}}{\Delta T} \tag{2.24}$$

$$\left.\frac{\partial^2 T}{\partial x^2}\right]_{i,j} = \frac{T_{i+1,j}^{k} + T_{i-1,j}^{k} - 2T_{i,j}^{k}}{\Delta x^2} \tag{2.25}$$

$$\left.\frac{\partial^2 T}{\partial y^2}\right]_{i,j} = \frac{T_{i+1,j}^{k} + T_{i-1,j}^{k} - 2T_{i,j}^{k}}{\Delta y^2} \tag{2.26}$$

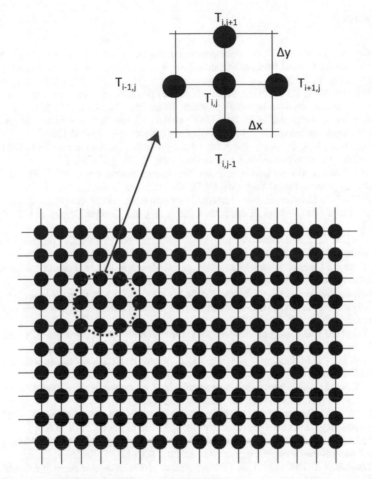

Fig. 2.22 Discretization of a surface into several points

where i and j indices represent the spatial directions and k represents temporal index.

Considering these equations and the heat transfer equation, the temperature in internal nodes at the increment time of k + 1 by explicit finite difference method is obtained by:

$$[T^{k+1}] = [A][T^k] + [q] \tag{2.27}$$

where [A] is the coefficient matrix and [q] is the energy addition to the system [44]. For the stability of the equation the time step should be defined correctly. Though both finite element and finite difference methods yield the same results in thermal analysis, the finite difference method is simpler to implement and facilitates the usage of materials properties variation.

References

1. Giudice, F., Missori, S., Sili, A.: Parameterized multipoint-line analytical modeling of a mobile heat source for thermal field prediction in laser beam welding. Int. J. Adv. Manuf. Technol. **112**(5), 1339–1358 (2021)
2. Dhinakaran, V., et al.: A review on the recent developments in modeling heat and material transfer characteristics during welding. Mater. Today Proc. **21**, 908–911 (2020)
3. Flint, T., et al.: Extension of the double-ellipsoidal heat source model to narrow-groove and keyhole weld configurations. J. Mater. Process. Technol. **246**, 123–135 (2017)
4. Bibby, M., Goldak, J., Shing, G.: A model for predicting the fusion and heat-affected zone sizes of deep penetration welds. Can. Metall. Q. **24**(1), 101–105 (1985)
5. Jia, X., et al.: A new method to estimate heat source parameters in gas metal arc welding simulation process. Fusion Eng. Des. **89**(1), 40–48 (2014)
6. Joshi, S., et al.: Characterization of material properties and heat source parameters in welding simulation of two overlapping beads on a substrate plate. Comput. Mater. Sci. **69**, 559–565 (2013)
7. Islam, M., et al.: Simulation-based numerical optimization of arc welding process for reduced distortion in welded structures. Finite Elem. Anal. Des. **84**, 54–64 (2014)
8. Gu, Y., et al.: Determination of parameters of double-ellipsoidal heat source model based on optimization method. Weld. World **63**(2), 365–376 (2019)
9. Hensel, J., Nitschke-Pagel, T., Dilger, K.: On the effects of austenite phase transformation on welding residual stresses in non-load carrying longitudinal welds. Weld. World **59**(2), 179–190 (2015)
10. Wu, D., et al.: Elucidation of the weld pool convection and keyhole formation mechanism in the keyhole plasma arc welding. Int. J. Heat Mass Transf. **131**, 920–931 (2019)
11. Schulze, G.: Die Metallurgie des Schweißens: Eisenwerkstoffe-Nichteisenmetallische Werkstoffe. Springer (2009)
12. Feng, Y., et al.: A 3-D lattice Boltzmann analysis of weld pool dynamic behaviors in plasma arc welding. Appl. Therm. Eng. **139**, 623–635 (2018)
13. Hildebrand, J.: Numerische Schweißsimulation-Bestimmung von Temperatur, Gefüge und Eigenspannung an Schweißverbindungen aus Stahl-und Glaswerkstoffen (2009)
14. Aissani, M., et al.: Three-dimensional simulation of 304L steel TIG welding process: contribution of the thermal flux. Appl. Therm. Eng. **89**, 822–832 (2015)
15. D'Ostuni, S., Leo, P., Casalino, G.: FEM simulation of dissimilar aluminum titanium fiber laser welding using 2D and 3D Gaussian heat sources. Metals **7**(8), 307 (2017)
16. Li, Y., et al.: An improved simulation of heat transfer and fluid flow in plasma arc welding with modified heat source model. Int. J. Therm. Sci. **64**, 93–104 (2013)
17. Farias, R., Teixeira, P., Vilarinho, L.: An efficient computational approach for heat source optimization in numerical simulations of arc welding processes. J. Constr. Steel Res. **176**, 106382 (2021)
18. Ahn, J., et al.: Determination of residual stresses in fibre laser welded AA2024-T3 T-joints by numerical simulation and neutron diffraction. Mater. Sci. Eng. A **712**, 685–703 (2018)
19. Bagchi, A., et al.: Numerical simulation and optimization in pulsed Nd: YAG laser welding of Hastelloy C-276 through Taguchi method and artificial neural network. Optik **146**, 80–89 (2017)
20. Chen, L., et al.: Numerical and experimental investigation on microstructure and residual stress of multi-pass hybrid laser-arc welded 316L steel. Mater. Des. **168**, 107653 (2019)
21. Zhan, X., et al.: The hourglass-like heat source model and its application for laser beam welding of 6 mm thickness 1060 steel. Int. J. Adv. Manuf. Technol. **88**(9–12), 2537–2546 (2017)
22. Busto, V., et al.: Thermal finite element modeling of the laser beam welding of tailor welded blanks through an equivalent volumetric heat source. Int. J. Adv. Manuf. Technol. 1–12 (2021)
23. Bonakdar, A., et al.: Finite element modeling of the electron beam welding of Inconel-713LC gas turbine blades. J. Manuf. Process. **26**, 339–354 (2017)

24. Li, Y.-J., et al.: Effects of welding parameters on weld shape and residual stresses in electron beam welded Ti_2AlNb alloy joints. Trans. Nonferrous Metals Soc. China **29**(1), 67–76 (2019)
25. Labeas, G., Moraitis, G., Katsiropoulos, C.V.: Optimization of laser transmission welding process for thermoplastic composite parts using thermo-mechanical simulation. J. Compos. Mater. **44**(1), 113–130 (2010)
26. Fraser, K.A., St-Georges, L., Kiss, L.I.: Optimization of friction stir welding tool advance speed via monte-carlo simulation of the friction stir welding process. Materials **7**(5), 3435–3452 (2014)
27. Grong, Ø., Sandnes, L., Berto, F.: A status report on the hybrid metal extrusion and bonding (HYB) process and its applications. Mater. Des. Process. Commun. **1**(2), e41 (2019)
28. Leoni, F., et al.: Rapid calculation of residual stresses in dissimilar S355-AA6082 butt welds. Materials **14**(21), 6644 (2021)
29. Pakkanen, J., Vallant, R., Kičin, M.: Experimental investigation and numerical simulation of resistance spot welding for residual stress evaluation of DP1000 steel. Weld. World **60**(3), 393–402 (2016)
30. Zhang, W.: Design and implementation of software for resistance welding process simulations. SAE Trans. 556–564 (2003)
31. Wan, X., Wang, Y., Zhang, P.: Numerical simulation on deformation and stress variation in resistance spot welding of dual-phase steel. Int. J. Adv. Manuf. Technol. **92**(5), 2619–2629 (2017)
32. Monnier, A., et al.: A coupled-field simulation of an electrical contact during resistance welding. In: Electrical Contacts-2006. Proceedings of the 52nd IEEE Holm Conference on Electrical Contacts. IEEE (2006)
33. Sheikhi, M., et al.: Thermal modeling of resistance spot welding and prediction of weld microstructure. Metall. Mater. Trans. A. **48**(11), 5415–5423 (2017)
34. Feujofack Kemda, B., et al.: Modeling of phase transformation kinetics in resistance spot welding and investigation of effect of post weld heat treatment on weld microstructure. Met. Mater. Int. **27**(5), 1205–1223 (2021)
35. Mikno, Z., Bartnik, Z.: Heating of electrodes during spot resistance welding in FEM calculations. Arch. Civil Mech. Eng. **16**, 86–100 (2016)
36. Zhang, W., Chergui, A., Nielsen, C.V.: Process simulation of resistance weld bonding and automotive light-weight materials. In: Proceedings of the 7th International Seminar on Advances in Resistance Welding. Busan, Korea (2012)
37. Zhang, W.: SORPAS-the professional software for simulation of resistance welding. DTU, Department of Manufacturing, Engineering and Management (2002)
38. Sajek, A.: Application of FEM simulation method in area of the dynamics of cooling AHSS steel with a complex hybrid welding process. Weld. World **63**(4), 1065–1073 (2019)
39. Michailov, V.: Erweiterte analytische Modelle fur die Berechnung der Temperaturfelder beim Schweissen. DVS Ber. **209**, 181–186 (2000)
40. Zhang, L., Kannengiesser, T.: Austenite grain growth and microstructure control in simulated heat affected zones of microalloyed HSLA steel. Mater. Sci. Eng. A **613**, 326–335 (2014)
41. Lu, Y., et al.: Numerical simulation of residual stresses in aluminum alloy welded joints. J. Manuf. Process. **50**, 380–393 (2020)
42. Wan, Y., et al.: Weld residual stresses in a thick plate considering back chipping: neutron diffraction, contour method and finite element simulation study. Mater. Sci. Eng. A **699**, 62–70 (2017)
43. Tchoumi, T., Peyraut, F., Bolot, R.: Influence of the welding speed on the distortion of thin stainless steel plates—numerical and experimental investigations in the framework of the food industry machines. J. Mater. Process. Technol. **229**, 216–229 (2016)
44. Stockman, T., et al.: A 3D finite difference thermal model tailored for additive manufacturing. Jom **71**(3), 1117–1126 (2019)

Chapter 3
Thermo-Metallurgical Modelling

Abstract Phase transformations in structural steels determine both thermal and mechanical properties of steel. Phase transformations depend on chemical composition, initial microstructure, maximum temperature and the cooling rate. During welding a gradient of temperature, occurring both temporally and spatially, may cause a variety of phases to co-exist and whose percentages are locally variable. Continuous-cooling-transformation (CCT) diagrams are reliable tools to calculate the phases percentages. The thermal and mechanical properties of steel for every location and every time can be determined based on the percentage of each phase. With the implementation of variable material properties in both thermal and mechanical analyses, a reliable and accurate prediction of joint performance is feasible during welding simulation. This chapter describes how material properties can be calculated and implemented in the simulation of a welded process and resultant welded joint.

3.1 Introduction

Metallurgical transformations determine the performance of a structure under various conditions and influence the development of residual stress and distortion during welding. These transformations are dependent on the thermal history and therefore the microstructure of a weld zone is usually not uniform during and after welding, depending on factors such as the base material, filler material and the process parameters. In fact, different regions of the weld zone may undergo vastly different metallurgical changes during welding. The basis of microstructural evolution is the conservation law of mass and energy and in welding simulation and metallurgical changes due to temperature variations are used to determine the material properties for use in thermal and mechanical modeling. The transient temperature fields change the microstructure and this has a direct effect on the thermal and mechanical properties of materials. Phase transformations cause an inhomogeneity in strain and stress in the material, as does the temperature distribution, while the resulting residual stresses and strains have no influence on the temperature field and on the phase transformations. It is this fact that enables the simulation of the phase transformation to be

SpringerBriefs in Computational Mechanics,
https://doi.org/10.1007/978-3-030-97910-2_3

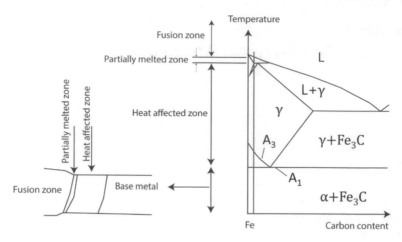

Fig. 3.1 A scheme of the weld region of a steel and the binary Fe–C phase diagram

carried out independently of the mechanical modeling process. The simulation of phase transitions is carried out in the thermal analysis and the results obtained are used in the mechanical modeling stage, which is fully decoupled from the thermal modeling.

However, not all the engineering alloys encounter phase transformations during welding processes, as is the case of some specific types of steels. In these cases, neglecting the phase transformation during simulation results in faster simulation process but this is not recommended for simulation of welding processes that include alloys that exhibit phase transformation, since it leads to a large loss of accuracy. Some of the currently available commercial software packages, such as SYSWELD, are capable of doing thermo-metallurgical-mechanical coupled process simultaneously, employing tools such as the binary phase diagrams which can provide a quantitative estimate of the phases that can be generated at different temperatures. However, these diagrams only apply under equilibrium conditions in which the heating and cooling rates are low. In fact, the rates of heating and cooling during welding are quite high and, in these conditions, such plots can struggle to provide accurate predictions. Figure 3.1 shows a schematic of the welded area of a steel and the applicable binary Fe–C phase diagram, identifying the temperature ranges corresponding to key areas of the weld. Further details regarding the phase evolution can be found in [1].

3.2   Phase Transformations in Steel

In structural steels, such as high strength low alloy steels, four different kinds of phase transformations can occur during heating and cooling. These are the austenite transformation during heating, diffusive transformations such as the ferrite and bainite transformation, and non-diffusive transformations such as martensite transformation during cooling. The critical transformation temperature are A_1 and A_3 for austenite transformation, B_s and B_f for starting and finishing temperatures for bainite transformation, and M_s and M_f for starting and finishing temperatures for martensite transformation [2]. During fusion welding processes, fusion and subsequent solidification establishes a bond between the two materials. If a filler material is used (either homogeneously or heterogeneously), a new material is added to the weld region which, in the case of heterogenous welding, has different chemical composition and material properties. In those cases, the material properties in the melting zone differ significantly from the base material and precise determination of the fusion boundary is crucial.

Phase transformations are accompanied by variation in physical and mechanical characteristics. Physical properties such as specific volume, heat capacity, and thermal conductivity and mechanical properties such as elastic modulus, yield strength, and thermal expansion coefficient can be determined by knowing the volume fraction of constituent phases at every location and for every time increment. The volume fraction of the constituent phases can be obtained using equations based on thermodynamic and kinetic concepts. These equations are either a function of both temperature and time for diffusion-based phase transformations or are solely a function of temperature for a non-diffusive transformation. During simulation, the material properties of each element can be calculated and updated for each time increment by knowing the volume fraction of the constituent phases and using the rule of mixtures. However, the process of calculating all material properties is very time intensive and, in practice, only a limited set of material properties are calculated and constantly updated during the simulation. The literature shows examples where some materials' properties such as density, heat capacity, thermal conductivity, Young's modulus and Poisson's ratio were considered independently from the phase transformation [3]. In another study, changes of materials properties due to phase transformation were not considered in thermal modelling [2]. In the mechanical analysis, the Young's modulus and the yield strength of the constituent phases and the thermal expansion coefficient and Poisson's ratio of the base material have been taken into account. Their dependencies on temperature were also taken into account. The temperature gradients in the weld and HAZ are higher than the rest of the workpiece and therefore, during thermal-metallurgical simulations, these regions should be finely discretized in a manner suitable to accurately capture the variation of materials' properties.

For some specific materials, such as non-alloyed structural steels, modelling can assume an homogenous condition since the effect of phase transformation is negligible when compared with the shrinkage caused by cooling from 500 °C to ambient temperature [3, 4]. In contrast, other cases exist where the transformation-induced changes are very significant and cannot be disregarded, for example when the phase transformation depends not only on the temperature but also on the cooling rate [4]. In addition, for high heating rates, transformation temperatures such as A_{C1} or A_{C3} are highly dependent on heating rate [3]. According to continuous cooling transformation (CCT) diagram, the variables that are used for prediction of phase transformation are the maximum temperature, the period of time spent above A_{C1} and the time spent between 800–500 °C during cooling [3]. In single pass welds, the time in which the different areas are held in austenitizing temperature are similar and therefore only the peak temperature can be used for calculation of the austenite fraction [3]. The time of austenitizing can also influence the grain size and consequently the yield stress of the austenite, the effect of which on the simulation results can be neglected. In fact, the grain size of austenite is especially important when it affects the subsequent phase transformations during cooling. For instance, it is reported that M_s temperature is a function of prior austenite grain size [5]. As it will be shown later in this chapter, the strains caused by martensite transformation are influenced by M_s, and therefore the grain size of austenite can be an important factor in welding simulation. In this case, both temperature and time of austenitizing are to be considered.

The solid–liquid phase transformation is usually not considered in welding simulations because the temperature at which this transformation occurs is sufficiently high to make the yield point zero. Therefore, neglecting this phase transformation in the mechanical analysis does not compromise the accuracy of the results. In a thermal analysis, the liquid phase in the melt pool promotes heat transfer by convection for which an artificial heat conductivity can be defined. The latent heat due to solid–liquid phase transformation can be incorporated into the model by changing the specific heat in solid and liquid states or by defining a new specific heat in the melting temperature range [6].

3.3 Effect of Phase Transformation on Residual Stress

Austenite transformations can influence the residual stress during welding, since an hindered volume expansion promotes compressive residual stresses [7]. This compressive stress is mainly caused by a difference in the specific volume of austenite and of the transformed phases (ferrite, bainite, martensite). The development of this compressive residual stress ends when the phase transformation is complete but with further cooling, residual tensile stresses begin to build up through rapid cooling (residual thermal stresses) and are superimposed on the previously formed compressive stresses. When the austenite cools slowly, it is transformed into ferrite and pearlite, a process which occurs only at high temperatures. The compressive residual stresses are thus low due to a low yield strength at high temperature, as shown in Fig. 3.2 by Curve 1 (the red lines show the limit on residual stresses because the residual stresses cannot exceed the yield strength). After the phase transformation is complete, the cool down process allows the thermal residual stresses to develop, which are superimposed over the previously developed compressive residual stresses. As the compressive residual stresses have low intensity, a total high residual tensile stress is developed. When the phase transformation occurs at lower temperature (by higher cooling rate which leads to the formation of martensite or bainite), a higher residual compressive stress is developed due to a higher yield stress at lower temperature. If the newly developed phases were to cool any further, thermal residual tensile stresses would arise. As the previously developed compressive residual stresses were

Fig. 3.2 The interaction of the compressive residual stresses due to phase transformation and the thermal residual stresses due to hindered shrinkage during cooling [7]

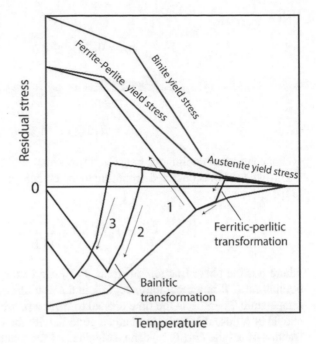

high, the resultant total residual stress would be compressive (curves 2 and 3 in Fig. 3.2). This process explains why a higher compressive residual stress occurs when a low transformation temperature (LTT) material is used as the filler material in the weld zone [8]. Please note that these curves are qualitative and only consider the volume change due to phase transformation while the transformation induced plasticity is neglected. Moreover, other factors such as strain or stress of austenite may also change M_s [9].

3.4 Calculation of Phase Percentage

Phase transformations may be diffusion controlled or non-diffusion controlled. For diffusion-controlled transformations, such as ferrite and bainite transformations, there are semi-empirical formulae which can predict the phase percentage by using CCT curves and can be implemented in FEM software. Some thermodynamic equations have also been developed to predict the phase fraction at different temperatures [2]. In structural steels, the austenitic phase formed during heating will be transformed to ferrite, pearlite, bainite, and martensite during cooling. Therefore, it is first important to determine the available austenite phase that forms upon heating. Only the knowledge of the local peak temperature (T_{max}) can provide a good approximation for estimating the percentage by volume of the austenite phase formed during heating. The following equations give the fraction of the austenite at each temperature

[10]:

$$p_\gamma = 0, T_{max} \leq A_{c1}; \tag{3.1}$$

$$p_\gamma = \frac{T_{max} - A_{c1}}{A_{c3} - A_{c1}}, A_{c1} \leq T_{max} \leq A_{c3}; \tag{3.2}$$

$$p_\gamma = 1, T_{max} > A_{c3}; \tag{3.3}$$

For continuous cooling the transformation kinetics of diffusion based transformations (like ferrite) is obtained according to the models proposed by Leblond-Devaux [11, 12]:

$$\frac{dp}{dt} = f(\dot{T}).\frac{p_{eq}(T) - p(T)}{\tau(T)} \tag{3.4}$$

where p is the phase fraction, t is time, $f(\dot{T})$ is a factor that takes into account the cooling rate, T is temperature, $p_{eq}(T)$ is the equilibrium content of the phase at temperature T, and τ is the time needed to form a phase component. These values should be adjusted in order to obtain a good fit with the CCT diagram. The volume fraction of bainite ($p_b(T, t)$) can be obtained by the kinetics model of Machnienko, described by the following equation [13]:

$$p_b(T, t) = \left(p_b^\% p_\gamma\right)\left(1 - \exp\left(-K_\gamma \frac{B_s - T}{B_s - B_f}\right)\right), B_s < T < B_f \tag{3.5}$$

where $p_b^\%$ is the final fraction of bainite obtained from CCT diagram, K_γ is a constant, and B_s and B_f are the starting and finishing temperatures of bainite formation, respectively. The phenomenological model for diffusion-less phase transformation (martensite transformation) is proposed by Koistinen-Marburger and is obtained from [11, 14]:

$$p_M(T) = \left(p_M^\% p_\gamma\right)(1 - \exp(-K_M.(M_s - T))), M_s < T < M_f \tag{3.6}$$

where $p_M(T)$ is the volume fraction of martensite, $p_M^\%$ is the final fraction of martensite obtained from CCT diagram, K_m is a coefficient, and M_s is the starting temperature for martensite formation. The value of K_m is dependent on the chemical composition of the steel. After evaluating the phase fraction at each time increment, the mechanical properties (yield stress and Young's modulus) can be estimated with the aid of the rule of mixtures.

$$M(T, t) = \sum p_i(T, t)M_i(T) \tag{3.7}$$

where $M(T, t)$ is the yield stress or Young's modulus, $p_i(T, t)$ is the volume fraction of phase 'i' and $M_i(T)$ is the corresponding properties of each phase 'i' at temperature 'T'. Thermal-physical properties can also be obtained by the mixture rule to be used in thermal analysis [10]. In the mechanical model, the total strain in each increment $(d\varepsilon^{th+\Delta V})$ which is sum of thermal strain $(d\varepsilon^{th})$ and phase transformation strain $(d\varepsilon^{\Delta V})$, is obtained according to the following equation [13]:

$$d\varepsilon^{th+\Delta V} = d\varepsilon^{th} + d\varepsilon^{\Delta V} = \sum_i p_i \alpha_i dT - sign(dT) \sum_i \varepsilon_i^{\Delta V*} dp_i \qquad (3.8)$$

where α_i is the thermal expansion coefficient of phase 'i', $\varepsilon_i^{\Delta V*}$ is the full volumetric change strain of phase 'i' caused by phase transformations obtained by dillatometric curves, and $sign(dT)$ is a sign function which is equal to $+1$ and -1 during heating and cooling, respectively. $\varepsilon_i^{\Delta V*}$ has a very significant effect on the residual stresses formed in the weld zone. The welding residual stresses in the longitudinal and transverse direction, taking into account the change in volume resulting from the phase transformation from austenite to martensite and without taking this effect into account, are shown in Fig. 3.3 for a P91 steel [15].

The transformation induced plasticity (TRIP) strain which results from the martensite transformation is obtained from [16]:

$$d\varepsilon_{trip} = 3k(1 - p_M).\Delta p_M.s \qquad (3.9)$$

where s is the deviatoric stress and Δp_M is the increment of martensite proportion. TRIP is referred to a micro plastic deformation caused by the superposition of an

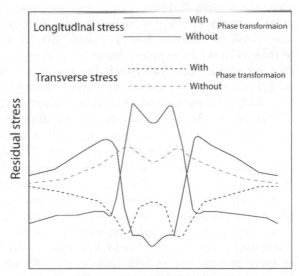

Fig. 3.3 The welding residual stresses in both longitudinal and transverse directions when considering the volume change resulted from phase transformation of austenite to martensite and when not considering this effect [15]. Adopted with permission from Elsevier

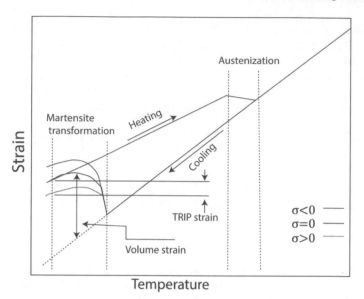

Fig. 3.4 The evolution of the TRIP strain, thermal strain and volumetric during martensite phase transformation [17]. Adopted with permission from Elsevier

external load and the internal stresses caused by the difference of specific volume of the parent phase (austenite) and the transformed phase (bainite or martensite) [17]. The evolution of the TRIP strain, thermal strain and volumetric strain is described in Fig. 3.4. During the heating process, a thermal strain is created due to the inherent thermal expansion coefficient of the materials and a volumetric strain arises due to phase transformation of martensite to austenite. However, the most important strains are those ones which develop during cooling. During cooling of austenite, a thermal strain builds up in the material and by reaching the M_s temperature, martensite begins to form and a shrinkage occurs due to this phase transformation. In the presence of an applied load, a transformation induced strain will appear and contributes to the total plastic strain. TRIP strain can moderate the longitudinal residual stress in multi-pass welding but has little effect on transverse stress [17]. In single-pass welding by electron beam welding, neglecting the TRIP strain in mechanical analysis is reported to increase the accuracy in the prediction of longitudinal residual stress but it over-predicts the transverse compressive stress [18]. Thermo-metallurgical models which neglect the TRIP strain, overpredict the compressive stresses which are developed in the weld zone by martensite transformation in multi-pass welding [19].

The implementation of material properties obtained by the equations related to phase transformation in commercial finite element software such as Abaqus can be performed in diverse manners, including the use of user-defined subroutines. For instance, the user subroutine USFLD can be used to define the properties of materials as a function of field variables [20]. The process of materials properties implementation in USFLD is shown in Fig. 3.5. The incremental thermal strain can be

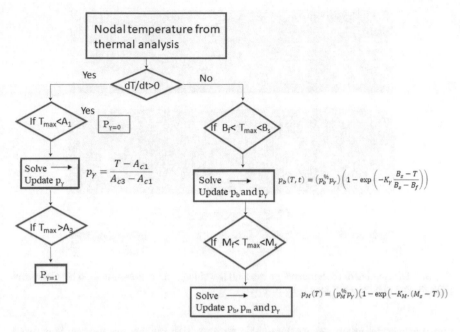

Fig. 3.5 Implementation of material properties in USFDL [20]. Adopted with permission from Elsevier

implemented in UEXPAN user-subroutine as a function of temperature, predefined field variables, and state variables.

3.5 Temper and Creep

Tempering is another phenomenon which may occur during welding of steels where martensite is formed. This transformation may occur during multi-pass welding or during a post-weld heat treatment and its occurrence lowers the residual stresses. The progress of this phenomenon is dependent on both temperature and time, though the temperature is more important [21]. Therefore, some researchers only consider the temperature to predict the tempered portion after welding and post-weld heat treatment [22]. Figure 3.6 shows the percentage of the tempered phase after welding and after the post-weld heat treatment. When the material is tempered, its yield stress decreases, and the residual elastic strain is relaxed when the yield stress of the tempered material becomes smaller than the residual stress. This leads to a reduction of the residual stress during a post-weld heat treatment.

The phenomenon of creep occurs at stresses below the yielding point and is dependent on the temperature, time and the local stress levels. Creep is a visco-plastic strain which can be included when modelling post weld heat treatments. The plastic

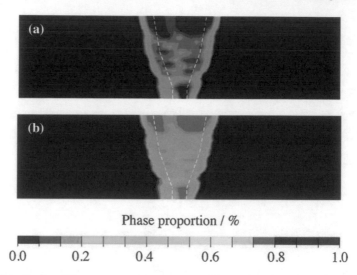

Phase proportion / %

0.0   0.2   0.4   0.6   0.8   1.0

Fig. 3.6 Distribution of the tempered phase, **a** after welding, and **b** after post-weld heat treatment [22]. Reprinted with permission from Elsevier

strain caused by creep lowers the elastic strain which means that the residual stress level is decreased. Please note that creep differs from the tempering effect, as stress relaxation can occur at stresses lower than the local yield stress and therefore a wider region is covered by this mechanism. It is shown that when the creep phenomenon during post-weld heat treatment is taken into account in the simulation, more accurate results can be obtained [22].

Residual stresses are highly dependent on the cooling rate, the temperature at which the phase transformation occurs, the yield strength at the temperature of phase transformation and the temperature gradient present during the cooling process after phase transformation is completed. The cooling rate is dependent on the heat input of welding and when the heat input is low, the cooling rate is high and the phase transformation occurs at low temperature, leading to a residual compressive stress. In contrast, when the heat input is increased, the phase transformation occurs at higher temperature and, as mentioned before, large residual tensile stresses are developed. However, by further increasing the heat input, the temperature of phase transformation will not increase. Instead, the cooling rate after phase transformation is decreased and the residual tensile stress is reduced.

References

1. Kou, S.: Welding Metallurgy, vol. 431, no. 446. New Jersey, USA, pp. 223–225 (2003)
2. Rong, Y., et al.: Residual stress modelling in laser welding marine steel EH36 considering a thermodynamics-based solid phase transformation. Int. J. Mech. Sci. **146**, 180–190 (2018)

3. Knoedel, P., Gkatzogiannis, S., Ummenhofer, T.: Practical aspects of welding residual stress simulation. J. Constr. Steel Res. **132**, 83–96 (2017)
4. Acevedo, C., Drezet, J.-M., Nussbaumer, A.: Numerical modelling and experimental investigation on welding residual stresses in large-scale tubular K-joints. Fatigue Fract. Eng. Mater. Struct. **36**(2), 177–185 (2013)
5. Heinze, C., et al.: Dependency of martensite start temperature on prior austenite grain size and its influence on welding-induced residual stresses. Comput. Mater. Sci. **69**, 251–260 (2013)
6. Promoppatum, P., et al.: A comprehensive comparison of the analytical and numerical prediction of the thermal history and solidification microstructure of Inconel 718 products made by laser powder-bed fusion. Engineering **3**(5), 685–694 (2017)
7. Hensel, J., Nitschke-Pagel, T., Dilger, K.: On the effects of austenite phase transformation on welding residual stresses in non-load carrying longitudinal welds. Weld. World **59**(2), 179–190 (2015)
8. Feng, Z., et al.: Transformation temperatures, mechanical properties and residual stress of two low-transformation-temperature weld metals. Sci. Technol. Weld. Joining **26**(2), 144–152 (2021)
9. Maalekian, M., Kozeschnik, E.: Modeling the effect of stress and plastic strain on martensite transformation. In: Materials Science Forum. Trans Tech Publ. (2010)
10. Hu, M., et al.: A new weld material model used in welding analysis of narrow gap thick-walled welded rotor. J. Manuf. Process. **34**, 614–624 (2018)
11. Neubert, S.: Simulationsgestützte Einflussanalyse der Eigenspannungs-und Verzugsausbildung beim Schweißen mit artgleichen und nichtartgleichen Zusatzwerkstoffen. Technische Universitaet Berlin (Germany) (2018)
12. Leblond, J., Devaux, J.: A new kinetic model for anisothermal metallurgical transformations in steels including effect of austenite grain size. Acta Metall. **32**(1), 137–146 (1984)
13. Piekarska, W., Kubiak, M., Saternus, Z.: Numerical modelling of thermal and structural strain in laser welding process. Arch. Metall. Mater. **57**, 1220–1227 (2012)
14. Krauss, G.: Steels: heat treatment and processing principles. ASM Int. **1990**, 497 (1990)
15. Suman, S., Biswas, P.: Comparative study on SAW welding induced distortion and residual stresses of CSEF steel considering solid state phase transformation and preheating. J. Manuf. Process. **51**, 19–30 (2020)
16. Rong, Y., et al.: Laser penetration welding of ship steel EH36: a new heat source and application to predict residual stress considering martensite phase transformation. Mat. Struct. **61**, 256–267 (2018)
17. Jiang, W., et al.: Effects of low-temperature transformation and transformation-induced plasticity on weld residual stresses: numerical study and neutron diffraction measurement. Mater. Des. **147**, 65–79 (2018)
18. Vasileiou, A.N., et al.: The impact of transformation plasticity on the electron beam welding of thick-section ferritic steel components. Nucl. Eng. Des. **323**, 309–316 (2017)
19. Chen, W., et al.: Thermo-mechanical-metallurgical modeling and validation for ferritic steel weldments. J. Constr. Steel Res. **166**, 105948 (2020)
20. Ghafouri, M., et al.: Finite element simulation of welding distortions in ultra-high strength steel S960 MC including comprehensive thermal and solid-state phase transformation models. Eng. Struct. **219**, 110804 (2020)
21. Mur, F.G., Rodriguez, D., Planell, J.: Influence of tempering temperature and time on the α'-Ti-6Al-4V martensite. J. Alloy. Compd. **234**(2), 287–289 (1996)
22. Ren, S., et al.: Finite element analysis of residual stress in 2.25 Cr-1Mo steel pipe during welding and heat treatment process. J. Manuf. Proc. **47**, 110–118 (2019)

Chapter 4
Thermomechanical Analysis in Welding

Abstract Mechanical analysis of welded joint is performed to calculate the residual stress and distortion generated during the welding process and cooling. Thermal elastic–plastic FEM analysis for doing so is usually performed using static implicit or quasi-static explicit methods. The strategies which can be employed to reduce the calculation time are described in this chapter. It is also explained how a mechanical analysis can be performed in a large-scale structure using approaches such as the inherent strain method. The effect of phase transformation in mechanical analysis is also discussed. Considering plasticization behavior of material and phenomena such as creep and annealing on mechanical analysis can increase the accuracy of simulation and in this chapter it is explained how one can implement these considerations in mechanical analysis. Finally, the effect of these phenomena on residual stress during welding are also addressed.

4.1 Residual Stress and Distortion

Residual stresses are, by definition, internal stresses and are mechanically in equilibrium and are formed without any applied external load. Knowledge about the type, amount, and distribution of residual stresses can help to predict structure performance under various loading conditions. During loading, residual stresses can influence the failure of the structure especially when they are aligned with the applied loads. Residual stresses can also influence stress corrosion cracking [1], cold cracking [2] and fatigue strength [3]. The residual stresses raised during welding are commonly termed as "welding residual stresses" and are formed due to a heterogenous plastic deformation field, which is caused by a local thermal source used in the welding process [4]. The existence of plastic deformation in every location is incompatible with its surroundings and occurs at the macroscopic, microscopic and submicroscopic levels in the weld, HAZ, and base material [5]. Residual stresses, when aligned in the loading direction, cause earlier plasticization and reduction of the local strength.

Residual stresses can be classified into four main groups:

1. Thermal, which are developed during cooling through differential deformation in the cross section, resulting from thermal expansion;
2. Deformation, which result from an inhomogeneous plastic deformation field from an external loading;
3. Phase transformation, caused by uneven phase transformation and a volume change;
4. Precipitation, a distortion stemming from the occurrence of precipitation in the crystal structure.

The sources of residual stresses in the welding region are shrinkage, rapid cooling and phase transformation [6]. Residual stresses due to shrinkage are always tensile in both transverse and longitudinal directions. To establish the equilibrium in a transverse direction, the compressive residual stresses are formed away from the weld in the HAZ and the base material. The mechanisms that underly the formation of residual stresses due to rapid cooling are similar to those formed due to shrinkage, occurring when the cooling rate is high enough to cause a large temperature gradient and severe constraints. In the thickness direction of thin sheets, in which the temperature distribution is uniform, this kind of residual stresses do not exist. The temperature around the weld zone is lower and therefore serves as a constraint in longitudinal direction. Perpendicularly to the weld, the boundary condition of the weld and the material of the weld both determine the residual stress [7]. In addition to the material, thickness and boundary conditions, the welding process and its parameters influence the residual stress. Figure 4.1 shows the residual stresses caused by shrinkage and

Fig. 4.1 Development of **a** longitudinal, and **b** transverse residual stresses due to shrinkage and rapid cooling and their superimposition along the longitudinal direction (x-direction) and the transverse direction (y-direction) for an Al–Si–Mg alloy [8]

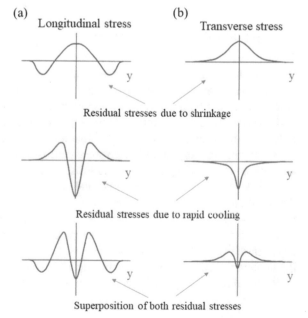

rapid cooling and their superimposition in both longitudinal and transverse directions for an Al-Si-Mg alloy [8].

A structure with large global stiffness is highly constrained. When the edge of the weld has no constraint, the composition of the weld is important in determination of the perpendicular residual stress [7]. In general, single-layer welds without an edge constraint develop lower residual stresses in the transverse direction. Hindered longitudinal shrinkage due to reaction forces leads to the reduction of the plastic deformation and develop residual stresses in the direction of the weld. The forces caused by the residual stresses in the structure are in balance and therefore the tension in the middle of the weld seam is balanced by compression at the end of the weld seam. Transversely to the weld seam, the residual stresses due to shrinkage reduce from the center to the edges. Generally, residual stresses in the thickness direction arise from different cooling rates along the thickness and are mostly formed in sufficiently thick sheets or in multilayer welding.

The determination of residual stress via numerical simulation requires a thermal analysis, used to first calculate the spatial and temporal temperature distribution and then coupled with a mechanical analysis to calculate the resulting stresses. Distortions are usually coupled with residual stresses and occur in the longitudinal, transverse and angular directions or any combination of these [5]. In general, a FEM simulation for mechanical analysis requires a set of governing equation, the concepts of which are described below.

A change in temperature in a metal bar is always associated with a change in its length, which is dependent on the modulus of elasticity. Any increase in temperature causes the interatomic distance to increase which is equivalent to a change in the lattice parameter. The elastic strain (ε_{el}) is related to the lattice spacing change according to the following equation:

$$\varepsilon_{el} = \frac{d - d_0}{d_0} \qquad (4.1)$$

where d is the lattice spacing of the specimen when is under stress and d_0 is the lattice spacing of the specimen when it is free of stress. This elastic strain is proportional to the temperature change according to the following equation:

$$\varepsilon_{el} = \alpha \Delta T \qquad (4.2)$$

where α is the coefficient of thermal expansion, which can be temperature dependent. If a metal is heated evenly and no constraint exists in the way of expansion, the atomic distance increases without inducing any interaction load. However, in the presence of constrains, residual stresses arise during the expansion and contraction of the material. As mentioned before, the constraints can be mechanical in nature or can result from uneven heating. Uneven heating is the main mechanism behind the development of residual stresses during a welding process, and any mechanical constraints can enhance this effect. In most cases, there is an interest to calculate the residual stresses after cooling. The origin of these stresses is the plastic strain that occurs when the elastic limit of the material is exceeded. The plastic yielding occurs when:

$$f = \sqrt{\frac{3}{2} \cdot \{S_{ij}\}^T \cdot \{S_{ij}\}} - \sigma_f = 0 \qquad (4.3)$$

where s_{ij} is the deviatoric stress and σ_f is the yield stress. The local yield point itself depends on the local temperature and the yield point of metals and alloys decreases with increasing temperature. The numerical FE simulation offers the possibility to take into account the variation of the yield strength with the temperature and to increase the accuracy of the mechanical models. The residual stresses in each spatial direction ($\sigma_{res,i}$) are related to the elastic strain according to Hooke's law:

$$\sigma_{res,i} = \frac{E}{1+\nu} \varepsilon_i + \frac{\nu E}{(1+\nu)(1-2\nu)} \left(\varepsilon_x + \varepsilon_y + \varepsilon_z\right) \qquad (4.4)$$

where E is the elastic modulus, ν is the Poisson coefficient, and ε_i is the strain in the direction of i. the elastic modulus is also temperature-dependent and decreases with increasing the temperature.

Phase transformations during cooling and heating can also contribute to the development of residual stresses, since they induce strains, namely phase transformation strains. In fusion welding processes, there is an increase in volume due to melting and, conversely, when solidifying. This type of phase transition has no effect on the residual stress development since both melting and solidification take place at high temperatures where the tensile stress is almost zero. The yield point is only significant below a certain temperature and can promote the residual stresses. Thus, one can neglect the melting and solidification processes as well as any phase transformation that occurs at high temperatures, since the yield strength is low at high temperatures and no residual elastic strain can be induced. For this reason, only the phase transitions that take place at lower temperatures should be considered in a mechanical analysis.

4.2   Numerical Simulation for Mechanical Analysis

The main outputs of the mechanical analysis are deformation and geometry variations as well as the stress and plastic strain distribution. There are two main approaches to calculate residual stress during welding. In the decoupled thermomechanical analysis, an FE heat transfer analysis is carried out first, which can also include the phase change analysis. Then, a separate FE thermal-stress analysis will be performed which is based on continuum mechanics and only calculates the macro residual stress [9]. This decoupling has a negligible influence on the simulation accuracy, since the mechanical state of the material has minimal influence on the thermal properties and the phase transformation [5]. An alternative approach performs both thermal and mechanical calculations at the same time which is referred to as coupled thermomechanical analysis. In a coupled transient thermal static structural analysis, the heat transfer problem is solved first, with the temperature history being saved for each node. Then, structural nodal loads are calculated using the thermal expansion coefficient in static conditions through the nodal temperature history. The coupled models assume that thermal energy and mechanical energy are equivalent and thus these analyses start with the fundamental equation of thermo-mechanics [10]:

$$cq\dot{T} + \dot{q}_{i,i} = \dot{Q}_v - \frac{E\alpha T}{1 - 2v}\dot{\varepsilon}_{eii} + \xi\sigma_{dij}\dot{\varepsilon}_{vpij} \tag{4.5}$$

where $cq\dot{T}$ is the heat stored energy per unit of time, $\dot{q}_{i,i}$ is the heat supplied or carried away (through surface), $\dot{Q}_v$ is the heat released or consumed per unit of time (related to volume), $\frac{E\alpha T}{1-2v}\dot{\varepsilon}_{eii}$ is the energy released from elastic deformation and $\xi\sigma_{dij}\dot{\varepsilon}_{vpij}$ is the energy due to viscoplastic deformation. A part of the energy used for plastic deformation is stored as microstructural changes and therefore a factor of ξ is used to state how much of this energy is converted to heat. This factor, in most cases, is close to 1 showing that almost all the energy for plastic deformation is converted towards heat. Coupling between thermal and mechanical fields can be ignored in welding as the heat is supplied solely from an outside source.

In a decoupled thermo mechanical analysis, two models, thermal and mechanical, are solved separately and sequentially. Usually, the mechanical analysis consumes more resources than the thermal analysis, due to the higher degree of freedom of the problem to be solved [11]. The stress distribution is calculated using FE methods, following the known strain components. Prior to each mechanical modelling step, the temperature distribution obtained from thermal analysis is loaded as a predefined field and the material properties of each node are suitably updated. Thermo-elastic–plastic (TEP) material behavior is described by the method of incremental strain change [12], where the total strain is obtained from [5, 12]:

$$\varepsilon = \varepsilon_{in} + \varepsilon_{el} + \varepsilon_{pl} + \varepsilon_{th} + \varepsilon_{tr} + \varepsilon_{tp} + \varepsilon_{cr} \tag{4.6}$$

In which ε_{in}, ε_{el}, ε_{pl}, ε_{th}, ε_{tr}, ε_{tp}, and ε_{cr} are initial, elastic, plastic, thermal, transformation, transformation plastic, and creep strains, respectively. For an isotropic material with an elastic modulus of E, a Poissons ratio of v, shear modulus of G, the strain in 3D in the elastic region (ε_{el}) is obtained from:

$$\varepsilon_{el} = [D]^{-1}.\sigma \tag{4.7}$$

where $[D]$ is the stiffness matrix. The plastic strain is derived from:

$$d\varepsilon_{pl} = \lambda_p.\frac{\partial f}{\partial\sigma} \tag{4.8}$$

where λ_p is a multiplier dependent on the problem and f is obtained from Eq. 4.2. The thermal strain is obtained from:

$$\varepsilon_{th} = \alpha(T).(T - T_0) \tag{4.9}$$

where $\alpha(T)$ is the thermal expansion coefficient which is dependent to temperature, T is the temperature and T_0 is the reference temperature. For the stress and strain field simulation, it is essential to determine the dependency of Poisson's ratio, Young's modulus, thermal expansion coefficient, and yield limit to the temperature [13]. There are two numerical approaches suitable to solve the mechanical models; static implicit and quasi-static explicit. In the following paragraphs, these two approaches are described in more detail.

4.2.1 Implicit Approach

The numerical simulation for mechanical modeling can be carried out in the static as well as in the dynamic states. In the static approach, no inertia effects are taken into account and only the load equilibrium is solved. A static approach for solving a mechanical model is applicable when the load frequency is less than ¼ of the natural frequency, so no inertia effects are present. In this way, the kinetic energy would be zero and the dynamic approaches (implicit dynamic and explicit dynamic) would give the same result as static approach. The dynamic approaches are mostly used when the inertia effect is not negligible but in the simulation of welding problems, the static approach is usually sufficient since inertia plays a minor role in the process. In a static analysis, when there are moving objects or an imposed displacement, it is important to properly define the constraints and limit the degree of freedom (DOF), otherwise the solving procedure will not converge.

For solving large-scale welding deformation problems using an implicit FEM analysis, the memory usage is high because there is a high order relationship between the arithmetic operations and the number of degrees of freedom [14]. Software packages such as JWRIAN [15] and SYSWELD are mostly based on an implicit method. The general equation for the static implicit FEM used in this process is:

$$K.\Delta U = \Delta F \tag{4.10}$$

where K is the stiffness matrix, ΔU is the increment of displacement and ΔF is the load increment. ΔF consists of both external load and thermal load. When 'n' nodes exist, the implementation of this equation in elastic region yields the following matrix:

Table 4.1 Velocity scaling method used in implicit approach [16]

Variable (unit)	Scaled value (ξ: Scale factor)
$T\,(\text{s})$	t/ξ
$\Delta t\,(\text{s})$	dt/ξ
$v\,(\text{mm/s})$	ξv
$\dot{Q}\,(\text{J/mm}^3\text{s})$	$\xi\dot{Q}$
$\dot{q}\,(\text{J/mm}^2\text{s})$	$\xi\dot{q}$
$\lambda\,(\text{J/mm s}^\circ\text{C})$	$\xi\lambda$
$\beta\,(\text{J/mm}^2\,\text{s}^\circ\text{C})$	$\xi\beta$
$C\,(\text{J/kg}^\circ\text{C})$	C
$\rho\,(\text{kg/mm}^3)$	ρ

$$
\begin{bmatrix} k & \cdots & \\ \vdots & \ddots & \vdots \\ & \cdots & k \end{bmatrix}
\begin{Bmatrix} \Delta u_1 \\ \vdots \\ \Delta u_{3n} \end{Bmatrix}
=
\begin{Bmatrix} \Delta f_1 \\ \vdots \\ \Delta f_{3n} \end{Bmatrix}
\tag{4.11}
$$

There are 3n possible displacements as every node can move in 3 directions. 'K' is the quasi-zero components of the stiffness matrix. The number of these components is proportional to n^2. The time increment used in an implicit analysis is determined by temperature increment. When an implicit analysis is used for heat conduction analysis, a velocity scaling factor can be used to shorten the analysis time. Accordingly, the heat flow rate and the materials characteristics related to time are scaled by a factor of ξ. Table 4.1 shows the values obtained by velocity scaling of specified physical quantities.

In a static implicit FEM used for welding mechanical analysis, several steps should be defined for applying the loads. A global stiffness equation should be solved for each load step which a very memory-intensive step for large-scale structures and with long computational times [17]. Welding problems possess high non-linearity especially around the weld line and the process of solving this non-linearity should be performed by iterative calculations such as Newton–Raphson method. The stiffness matrix of the whole structure is updated for each increment which is very time-consuming. Thus, an iterative substructure method (ISM) is often used to avoid frequent updating of the global stiffness matrix [11, 14]. With this method, the area near the moving heat source, which has a strong non-linearity, is solved separately from the remaining region (Region 'B' in Fig. 4.2a). The remaining region (region A-B in Fig. 4.2a) behaves elastically and an elastic stiffness matrix is assigned to the whole structure (K_A^e) initially, with its inverse being saved. The stiffness matrix of the nonlinear region (K_B^n) is updated in every temperature increment while the remaining region (A-B) is only solved and updated every N_{A-B} increment using the initially saved inverse matrix of K_A^e. Since the stiffness matrix is only updated incrementally for region 'B', the computational time can be shortened. The continuity of the displacement at the interface of the two regions can be directly maintained and

(a)   (b)

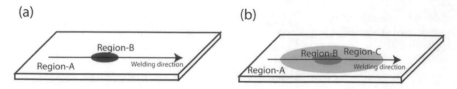

Fig. 4.2 Computation of stiffness and stress in **a** ISM method, and **b** i-ISM method

the continuity of the traction is maintained by the iterative correction process. For larger components, defined as those that consist of more than 1 million elements, one can introduce even more regions in order to shorten the calculation time (Fig. 4.2b). In this way, region C-B is treated like the region A-B in the previous example. Therefore, region A-C requires a lower number of stiffness matrix updates by a factor of $1/N_C$, in which N_C is the number of steps needed to update the stiffness matrix of region C-B. This method is named the inherent strain ISM (i-ISM) method. In the following paragraphs, it is described how the ISM method can be implemented during welding by a moving heat source.

As mentioned, in ISM one area around the welding zone exhibits strong nonlinear behavior while the remaining areas exhibit linear behavior or a small stiffness change (quasi-linear). Figure 4.3.a shows how the stiffness matrix is updated when the heat source is moving. The non-linear region moves as the torch moves and the nonlinear region thus changes from 'B' to 'B'' and the stiffness matrix is updated to nonlinear, while the stiffness matrix of the remaining region remains linear. The continuity at the boundary of these two regions should be maintained in the iterations. The stiffness matrix of region A' + B' can be maintained over the process as long as there is convergence in the solution and the initial stiffness matrix can be used until the end of the process. This reduces the calculation time considerably. The schematic of the procedure used to calculate the stress is shown in Fig. 4.3b. The stress is obtained by considering the continuity and balance of stress at the boundary. It is important to define the border between the linear and non-linear regions properly. Underestimating the extent of the non-linear region will compromise the accuracy of the results, although the computational time will decrease. The extent of the nonlinear area can be defined as a distance over which the accuracy of the results does not change (is not improved) and this area can be defined as a region whose maximum temperature is greater than a specified one. In this area, the temperature gradient is steep, and the local yield point is low enough to induce local plastic strains and residual stresses. The mesh size in the linear region is set to be coarser since it is linear in nature.

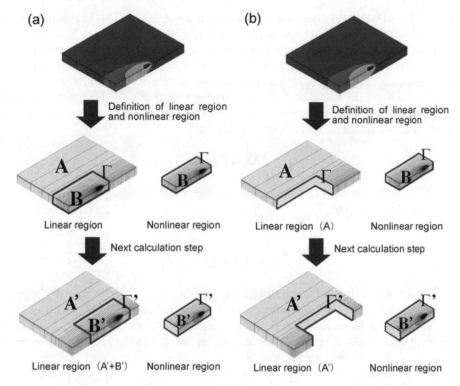

Fig. 4.3 Region division in the iterative substructure method for **a** constituting the stiffness matrix, and **b** stress calculation [18]. Reprinted with permission from Elsevier

4.2.2 Explicit Approach

The explicit method is suitable for modelling dynamic phenomena which take place in a short period of time [16]. However, it can be adapted for mechanical analysis of welding which in this case is named as a quasi-static explicit method. In explicit method, the time increment (time step) should be selected in a way to ensure that the stress wave propagation distance is smaller than the element size. The basic equation of dynamic explicit FEM is [17]:

$$[M]\{\ddot{U}\}_t + [C]\{\dot{U}\}_t + \sum_{e=1}^{Ne} \int [B]^T\{\sigma\}dV = \{F\}_t \qquad (4.12)$$

where $[M]$, $[C]$, $[B]$, and $\{\sigma\}$ are the mass matrix, damping matrix, strain–displacement matrix and stress vector, respectively. $\{\ddot{U}\}_t$, $\{\dot{U}\}_t$, and $\{F\}_t$ are acceleration, velocity, and load vector at time t. Ne is the number of elements and dV. is the volume of each element. In the simulation of welding, the scale of simulation is usually limited to the welding joint area. Idealized Explicit Finite Element Method

(IEFEM) can be used to perform simulations in larger scales [17]. In this method, the load is applied in several increments and the displacement in every increment is calculated based on the above equation until the static equilibrium state is reached in that load increment. Thereafter, a new load increment is applied and the procedure is repeated. For large scale problems, the amount of memory needed for an explicit analysis is generally less than that necessary for an implicit one [16], since in the explicit method, there is no need to perform convergence calculations. The time increment in the explicit analysis should then be:

$$\Delta t < L/c \tag{4.13}$$

where 'L' is the minimum size of an element and 'c' is the wave propagation speed which is obtained from:

$$c = \sqrt{\frac{E(1 - v)}{\rho(1 + v)(1 - 2v)}} \tag{4.14}$$

Some methods such as mass scaling, are commonly used to shorten the calculation time in an explicit analysis. By scaling the mass, 'c' is virtually reduced and therefore the step time is increased. The elastic modulus and Poisson ratio should not be changed because these parameters directly affect the calculation of the stiffness and stresses acting on the material. However, using mass scaling causes a degree of dynamic vibration which can be eliminated by considering a damping coefficient between 20–100% of the minimum angular frequency in the analysis [16]. By using an explicit method we can take the advantage of using several CPU or Graphic Processing Unit (GPU) cores in parallel, as the explicit equations can be solved independently for each node [14]. This parallelization can take the advantage of GPU to accelerate the computation time in explicit method. Figure 4.4 shows a comparison between the two approaches, static implicit and explicit (idealized and

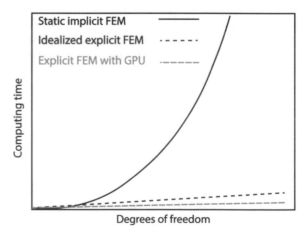

Fig. 4.4 Comparison between the two approaches, static implicit and explicit (idealized and using GPU) in terms of calculation time [17]. Adopted with permission from Elsevier

using GPU) in terms of calculation time. In a static implicit approach, the stiffness matrix is updated in each increment and, depending on the DOF of movement of each element, several elements of the stiffness matrix must be calculated and updated. In the explicit method, the mass matrix is updated in each increment which is a diagonal matrix and a reduced number of elements need to be updated. Hence, for a large structure an explicit analysis requires a lower calculation time.

Velocity scaling is another commonly used technique to reduce calculation time in a FEM explicit deformation analysis performed for quasi-static condition. In this method, strain velocity is accelerated. The temperature change rate can also be accelerated to decrease the calculation time. By using velocity scaling in a thermal elastic–plastic analysis, the mechanical characteristics of materials such as Young's modulus E, Poisson's ratio v, yield stress s, work-hardening coefficient H and thermal linear expansion coefficient cannot be scaled.

4.3 Thermal Elastic–Plastic FEM for Small Scale Structures

The thermal elastic–plastic (TEP) method can yield both the temperature and stress fields and is mostly suitable for analysis of small structures, since it is computational heavy [13]. In this way, the weld is considered as a transient nonlinear problem. Figure 4.5 shows the steps needed in the thermo elastic–plastic FEM analysis to determine the residual stress and deformation [19]. Small structures, which typically consist of a single weld line, can be analysed using this method. For larger structures, which are made up of multiple components connected by several weld lines,

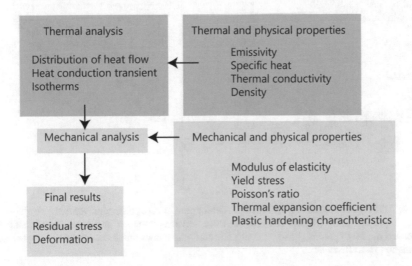

Fig. 4.5 The flowchart of TEP FEM

TEP analysis leads to computational times which are much higher and therefore this method is unpractical for these applications and rarely used in full structures. In those more complex cases, weld lines are first classified based on their geometry, dimension, and welding process. Instead of analysing all the weld lines simultaneously, only each classification will be analysed by the TEP method. Thereafter, the results of the TEP analysis are used as input for other approaches that are more efficient and thus can be used to perform a mechanical analysis of the entire structure. Two of these approaches are discussed in the next two subsections.

Before it can be used, a TEP analysis needs to be calibrated and validated experimentally. When TEP is used on a small yet complex structure, the validation of the analysis can be performed on a simplified specimen, dispensing tests on the real structure, since the real structure often consist of components that may have been prefabricated by complex processes. Figure 4.6 (left) shows the results of a calibrated model for prediction of distortion verified by experiment on a simple lap joint. In Fig. 4.6 (right) a real structure is shown, which is complex and simulated by the calibrated model to predict the distortion level. No experiment is performed to validate this result.

In a TEP FEM it is fundamental to consider the interaction of the materials properties with both the thermal and mechanical models [21], as shown in Fig. 4.7. The consideration of both physical and mechanical properties and their temperature dependency during the simulation can be decisive in the thermal and mechanical analysis. The effect of material properties will be discussed later in this chapter.

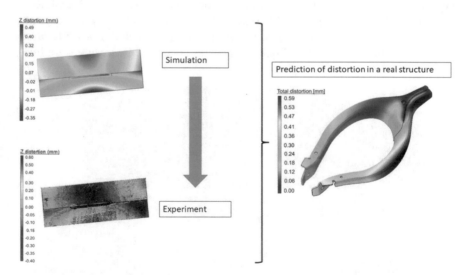

Fig. 4.6 (Left) The results of a calibrated model for prediction of distortion verified by experiment on a simple lap joint. (Right) The real structure which is complex and is simulated using the calibrated model to predict the distortion. The color map shows the distortion [20]. Adopted with permission from Elsevier

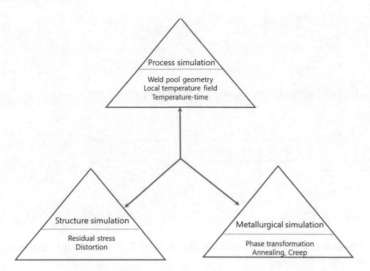

Fig. 4.7 Interaction of the materials properties with both thermal and mechanical models

4.4 Inherent Strain Method

As previously stated, if one wishes to calculate the residual stress and the distortion in a large and complex structure the use of TEP-FEM as the sole method is highly discouraged, since it is an extremely resource heavy and time-consuming process. An alternative approach for the determination of the distortion and residual stress for large and complex structures consists in the use of the inherent strain method. In this way, a TEP-FEM and an linear FEM analysis are used to calculate the residual stress and distortion [22]. The linear FEM analysis uses an inherent strain as initial strain. The inherent strain is obtained by using TEP-FEM method for each weld line and this considerably reduces the calculation time. In the linear analysis, the structure can be fully represented by shells, which further shortens the calculation time [13]. The level of this inherent strain is dependent on the peak temperature and the existing constraints [23]. In this approach, the inherent strain values calculated by TEP FEM simulation for each individual weld are used as an initial load in the mechanical linear FEM simulation in a global shell, allowing to calculate the distortion in large and complex structure [24]. Therefore, the time needed for calculation is drastically decreased, as the non-linear TEP FEM model does not need to be applied to the whole structure. This procedure is shown schematically in Fig. 4.8 [25].

The inherent strain calculated by TEP-FEM analysis is converted to either force or deformation for use in a linear FEM analysis. When a strong constraint exists, the inherent strain is transformed to stress and when there is a small restraint the inherent strain is transformed to deformation [22]. In the linear FEM model for calculation of transverse shrinkage and bending (angular distortion) the inherent deformation can be used when the restraints in transverse direction are negligible. For longitudinal shrinkage and bending, the inherent force or moment can be used for linear FEM

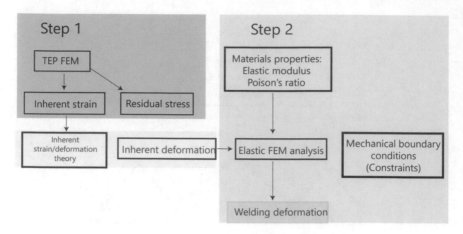

Fig. 4.8 The flowchart of calculation of inherent strain by TEP FEM (step 1) and using it in the subsequent linear FEM (step 2)

modelling, as the restraint in longitudinal direction is huge. Figure 4.9 shows the transverse shrinkage (δ_y^*) as well as the longitudinal force (F_L^*.) resultant from the longitudinal shrinkage calculated using the inherent strain components (ε_x^*. and ε_y^*). Shrinkage in one direction is always associated with shrinkage in the other direction, and these two are linked by Hooke's law. Figure 4.9c and d show both mentioned deformation and force in an element.

These concepts can be integrated into a FEM model, allowing to predict the residual stress and distortion [22]. The inherent strain is obtained from the following equation:

$$\varepsilon^* = \varepsilon - \varepsilon^e = \varepsilon^p + \varepsilon^{th} + \varepsilon^{cr} + \varepsilon^{tr} \tag{4.15}$$

ε^p, ε^{th}, ε^{cr}, and ε^{tr} are plastic strain, thermal strain, creep strain, and transformation strain, respectively. ε^* is considered as the inherent strain which contributes to the welding deformation. The inherent strain can be schematically explained as depicted in Fig. 4.10. When a material is stressed, a strain is induced and when a piece of the material is cut, one part of the strain is released which is elastic and associated to the residual stress. The component of strain that is irrecoverable after cutting is referred to as an inherent strain [27].

The inherent strain can also be obtained experimentally. When using the simulation results obtained by TEP FEM, it is important to verify those results with some experiments. The inherent strain values are uniform along the weld line except the beginning and the end of the joint. It is because the thermal condition at the beginning and the end of the joint is not quasi-static. In addition, the restraints at the beginning and at the end of the weld are different from those in the rest of the weld line.

The inherent strains are usually calculated in the transverse direction and in some cases in the longitudinal direction [24]. The deformations caused by inherent strain

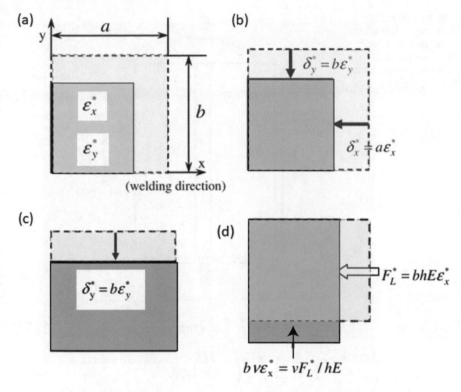

Fig. 4.9 **a** Shrinkages represented by inrent strain. **b** Shrinkage represented by inherent deformation. **c** Transverse shrinkage. **d** Longitudinal shrinkage [26]. Adopted with permission from Elsevier

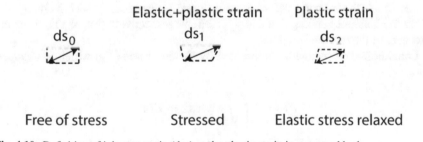

Fig. 4.10 Definition of inherent strain (ds$_2$) as the plastic strain in a stressed body

values are averaged through the thickness (z), perpendicularly to the weld direction (y-direction) according to the following equations [22]:

$$\delta_x^* = \frac{1}{h} \int \int \varepsilon_x^* dy dz \qquad (4.16)$$

Fig. 4.11 Schematic of shrinkage and bending resulted from inherent strains in transverse and longitudinal directions

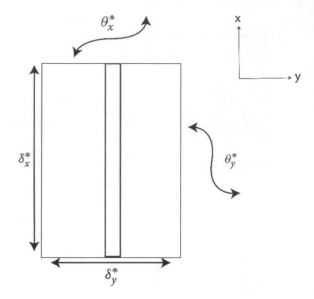

$$\delta_y^* = \frac{1}{h} \int \int \varepsilon_y^* dydz \tag{4.17}$$

$$\theta_y^* = \frac{12}{h^3} \int \int \left(z - \frac{h}{2}\right) \varepsilon_y^* dydz \tag{4.18}$$

$$\theta_x^* = \frac{12}{h^3} \int \int \left(z - \frac{h}{2}\right) \varepsilon_x^* dydz \tag{4.19}$$

δ_x^* and δ_y^* are the inherent deformation in x and y directions, respectively. θ_x^* and θ_y^* are the inherent bending in x and y directions, respectively (Fig. 4.11). 'h' is the thickness of the welded structure.

Longitudinal shrinkage is represented by Tendon force (F_T) which is obtained from:

$$F_T = E \int \int \varepsilon_x^* dydz = Eh\delta_L^* \tag{4.20}$$

This force is the driving force for buckling in thin plates whose resistance to buckling is low. The value of Tendon force in the beginning and in the end of the weld is different from the rest due to the free restrain effect. An average of this force along the longitudinal direction is calculated, disregarding the end and beginning, as shown in Fig. 4.12 [28]. This average value accounts for the total Tendon force that will be used as an inherent force in linear FEM model. In the linear FEM model, the longitudinal shrinkage, the transverse shrinkage, and the angular distortion are used as initial strains. Instead of the longitudinal shrinkage, the Tendon force can be used to represent the longitudinal shrinkage. When transferring the results

Fig. 4.12 Distribution of
tendon force along the
longitudinal direction of the
weld

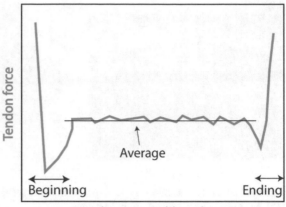

obtained from TEP FEM analysis into the global model, it is important to choose
the right boundary conditions and welding sequences. Away from the weld zone the
inherent strain is equal to the plastic strain, since the thermal strain and the phase
transformation induced strain are negligible [29]. In the linear FEM analysis, the
inherent deformation components are employed as mechanical loads (displacements)
[25].

The inherent strains can also be measured by experimental methods such as cutting
procedure [29]. In this way, the residual stress as well as the deformation in the
structured can be calculated using the experimentally measured inherent strains. The
basic equations of the linear FEM model are [29]:

$$[K]\{u\} = \{f^*\} \tag{4.21}$$

$$\{f^*\} = \int [B][D]\{\varepsilon^*\}dv \tag{4.22}$$

where $[K]$, $\{u\}$, and $\{f^*\}$ are the elastic stiffness matrix, the nodal displacement and
the inherent strain induced nodal force. dv is the volume of element and $[B]$ is the
corresponding matrix between the nodal displacement and strain in an element. The
displacement field obtained from the above equations will be used in the following
equations to calculate the total strain ($\{\varepsilon\}$.) and the residual stress ($\{\sigma\}$):

$$\{\varepsilon\} = [B]\{u\} \tag{4.23}$$

$$\{\sigma\} = [D^e](\{\varepsilon\} - \{\varepsilon^*\}) \tag{4.24}$$

where $[D^e]$ is the material elastic matrix.

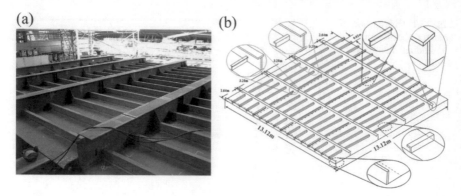

Fig. 4.13 **a** Plates and stiffeners of a ship panel structure. **b** The categorized weld lines in the panel [25]. Reprinted with permission from Elsevier

The inherent strain calculated using the TEP-FEM method is mainly used for deformation analysis by linear FEM. Figure 4.13a shows a car deck panel structure used in a large ship. It consists of plates and stiffeners which are welded together. Please note, however, that the welding conditions are different for different joints of the structure. Figure 4.13b shows that the welds can be categorized into six main groups. In this example, in order to determine the buckling of the structure during welding, a TEP-FEM analysis was first carried out for each type of weld in order to obtain the inherent strain. The simulation results for each category of weld were verified experimentally. The inherent strains obtained from this TEP FEM model were used in an elastic FEM model to measure the distortion by both small deformation theory and large deformation theory. Buckling of the structure was calculated without considering the bending in both directions and only transverse and longitudinal shrinkage were used for doing so.

When assembling the parts for the linear FEM analysis, it is necessary to position each part so that is initially free to move, and a tight contact relationship between the elements must be established. An interface element must then be defined between the parts to account for contact, slide and gap between the parts. After assembling and positioning the part, the stiffness in different directions is defined for each element. In this way, the constraining effects such as tack welds or fixturing can be accounted for in each interface. After completion of the weld at each interface, a very high stiffness value should be defined, which corresponds to the tightening of the mating parts [26]. This procedure highly influences the calculation results as shown in Fig. 4.14a and b. In Fig. 4.14a no gap correction was considered and in Fig. 4.14b this was corrected. This Fig. shows how defining a controlled gap between the stiffeners and the skin can reduce out-of-plane distortion. The value of stiffness defined for interface element between the stiffener and the skin controls this out of plane distortion.

(a)   (b)

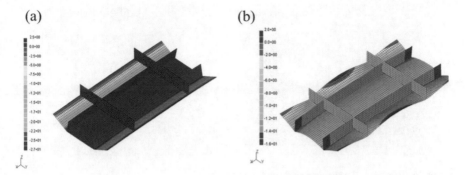

Fig. 4.14 The influence of gap correction in the calculation of distortion by inherent strain method, **a** without gap correction, and **b** with gap correction [26]. Reprinted with permission from Elsevier

4.5 Equivalent Thermal Strain Method

The inherent strain method is only applicable for heating or welding lines which are straight. In curved plates, the introduction of forces caused by inherent strain cannot establish an equilibrium and causes a rigid body movement [30]. The equivalent thermal strain method was effectively used to calculate the welding distortion during welding a pipe [23] and the distortion in other complex structures by inherent strain [30]. The concept of this method originates from the law that states that the equivalent thermal strain is equal to the product of an artificial thermal expansion (α) coefficient and an artificial temperature gradient through the thickness (ΔT) according to:

$$\varepsilon_{th}^{eq} = \alpha \Delta T \tag{4.25}$$

The inherent strain can be directly substituted in the artificial thermal expansion coefficient:

$$\alpha = \varepsilon^* \tag{4.26}$$

The artificial thermal coefficient is zero in areas away from the weld seam in which there is no intrinsic inherent strain. The temperature gradient is defined according to the shape of the inherent strain region around the weld line. This temperature gradient can be obtained by calculating the force and momentum according to two approaches (inherent strain, and equivalent thermal strain methods) and equating the obtained values. The inherent strain in the welding region is almost compressive and asymmetrical across the thickness (Fig. 4.15). The compressive force causes a shrinkage force in the transverse direction and its asymmetry causes a momentum (angular distortion) with respect to the neutral axis (N.A.). This shrinkage force (F) can be calculated by inherent strain value according to:

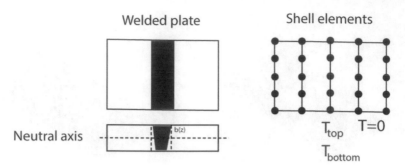

Fig. 4.15 Principle of imaginary node and temperature

$$F = E\varepsilon^* \int b(z)dz \tag{4.27}$$

where b(z) is defined based on the shape of the inherent strain distribution (see Fig. 4.15). The shrinkage force can also be calculated using the equivalent thermal strain method according to:

$$F = E\alpha \Delta T bh \tag{4.28}$$

where b is the extent of inherent strain and h is the height of it. Similar calculations can be performed for the determination of moment. By equating the forces obtained from both approaches as well as equating the momentum values, the artificial temperature on the top and bottom of the inherent strain region is obtained as:

$$T_{Top} = \frac{1}{A} \int_{-\frac{h}{2}}^{\frac{h}{2}} b(z).\left(1 - \frac{4}{h} \times z\right)dz \tag{4.29}$$

$$T_{Bottom} = \frac{1}{A} \int_{-\frac{h}{2}}^{\frac{h}{2}} b(z).\left(1 + \frac{4}{h} \times z\right)dz \tag{4.30}$$

The artificial temperature difference at the top and bottom is responsible for the angular distortion. Further modifications have been applied in the literature [31] to accurately define the artificial temperature and artificial thermal coefficient not only in the transverse direction, but also in the longitudinal one. A trapezoidal shape is mostly used for the inherent strain distribution but other shapes can also be used [30]. The element size is defined by the elements of nodes at which the equivalent forces are applied and is equal to the extent of HAZ [30].

The flowchart of equivalent thermal strain method is depicted in Fig. 4.16. First, a TEP FEM analysis is carried out to calculate the inherent strain in the weld zone.

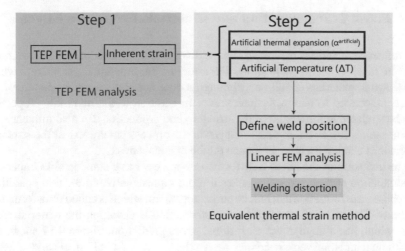

Fig. 4.16 Computation procedure for equivalent thermal strain method

Thereafter, the output of this simulation will be converted to an artificial thermal expansion coefficient and artificial temperature gradient to be used in the linear FEM analysis. The computing time in this procedure is much less than thermal mechanical FEM analysis, as it is a linear FEM. Solving this linear FEM involves two stages, of thermal and mechanical nature. First, the thermal model is solved by linear interpolation taking into account the calculated artificial temperatures. The temperature beyond the inherent strain region is made to be equal to zero. In the second step, the temperature distribution obtained by this thermal FEM analysis is loaded into the mechanical analysis as initial condition. The mechanical model is then solved by considering the artificial thermal expansion coefficient [31]. The value of artificial thermal expansion beyond the inherent strain region is equal to zero. The artificial temperature distribution, as representative of the inherent strain region, determines the bending moment with respect to the neutral axis, as well as the shrinkage level. The transverse distribution of artificial temperature represents the inherent strain component in the transverse direction. The transverse shrinkage and bending are obtained by assigning ε_y^* to α.

In addition to the angular and shrinkage distortion, longitudinal distortion can also be modelled by this method, as proposed in [31]. For doing so, the artificial thermal coefficient expansion should be first defined in both the transverse and longitudinal directions. That means that both transverse and longitudinal inherent strains must be calculated after TEP FEM simulation is carried out. However, whether the longitudinal expansion should be considered depends mainly on the structure type and the importance of the calculation of the longitudinal shrinkage.

4.6   Phase Transformation and Materials Properties Effects

The influence of various parameters on the diverse types of residual stress is summarized in Table 4.2. This table shows the effect of the individual parameters which are related to materials conditions, although it does not provide a clear picture of the effects associated to parameter interaction. In addition to the materials properties, variables related to the design of the structure and production can also influence the residual stress [5]. Even minimal changes in the process parameters or the structure dimensions can have a huge influence on the residual stress.

The inclusion of the phase transformation process has a considerable impact on the simulation accuracy since it makes it more representative of the real condition. The phase transformation can induce permanent strain which is called transformation plasticity. The other effect of phase transformation is changing the material properties which has a large effect on residual stress prediction. Figure 4.17 shows the longitudinal and transverse residual stress in the surface of a St37 steel welded with a double V-groove. Microstructural changes occur in the weld zone and HAZ (martensitic transformation) and the differences between the two curves (with and without microstructural change) are noticeable in these regions. In this regard, the proper choice of material properties by considering the phase transformations is critical for the accurate prediction of the residual stress [32]. Usually, phase transformations are predicted in the thermal analysis with the assumption that stress distribution has no effect on the phase transformation [9]. In other words, in the case of simulation problems, it is assumed that the phase transformations are only dependent on the temperature history and the material properties are therefore determined solely in the thermal analysis. The material properties obtained in the thermal analysis are then transferred and used in the mechanical analysis.

In some cases, it is important to consider the hardening effect in the material in the mechanical analysis. Both kinematic and isotropic hardening have been used to

Table 4.2 Effect of various parameters on different kinds of residual stresses [5]

Shrinkage residual stresses change with:							
$\sigma_{YS}\uparrow$	$d\sigma_{YS}/dT\uparrow$	$\alpha\uparrow$	$E\uparrow$	$\lambda\downarrow$	$dT/dX\uparrow$	$s\downarrow$	$t\uparrow$
Rapid cooling shrinkage residual stresses change with:							
$\sigma_{YS,\Delta T_{max}}\downarrow$	$\sigma_{YS}-$ $\sigma_{YS,\Delta T_{max}}\uparrow$	$\alpha\uparrow$	$E\uparrow$	$\lambda\downarrow$	$\Delta T_{max}\uparrow$	$dT/dt\uparrow$	$t\uparrow$
Phase transformation residual stresses change with:							
$\sigma_{YS,Tu}\uparrow$	$\sigma_{YS,Tu,min}-$ $\sigma_{YS,Tu,max}\uparrow$	$\Delta V_{tr}\uparrow$	$E\uparrow$	$\lambda\downarrow$	$M_s\downarrow$	$dT/dt\uparrow$	$s\downarrow$

σ_{YS} is yield stress, $\sigma_{YS,\Delta T_{max}}$ is yield stress at temperature wherein the temperature difference is maximum, $\sigma_{YS,Tu}$ is yield stress at temperature of phase transformation, α is thermal coefficient, λ is thermal conductivity, E is elastic modulus, s is the width of the seam weld, t is thickness, dT/dt is cooling rate, dT/dX is temperature gradient, M_s is the starting temperature of martensite, ΔV_{tr} is volume change due to phase transformation.

Fig. 4.17 The longitudinal and transverse residual stress in the surface of a St37 steel welded with a double V-groove either with or without considering the phase transformation [33]

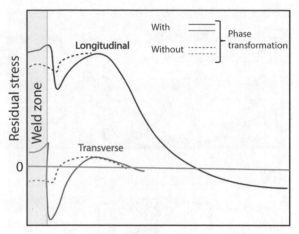

calculate the residual stress [18]. With multi-phase welding, the thermal effect of the subsequent layers on the previous layers must be taken into account. In multi-pass welding, a thermal cycle is experienced by each layer which schematically is shown in Fig. 4.18. The thermal history is dependent on several factors such as welding process, material properties, initial or ambient temperature, and the dimension of the weld. When the weld line is long enough, the temperature of each point diminishes to the initial temperature, i.e. the inter-pass temperature is not changing. When the weld line is short there is not enough time between the weld passes for cooling down and the temperature in every point raises after each pass. In these cases, the presence of a thermomechanical cyclic load requires the application of mixed isotropic-kinematic hardness laws in order to take into account both the Bauschinger effect and the cyclic hardening [34]. Figure 4.19 shows these models schematically. Isotropic hardening defines the evolution of the yield stress surface and is obtained from [34]:

$$\sigma^0 = \sigma_0 + Q_{inf}\left(1 - \exp^{-b\dot{\bar{\varepsilon}}_{pl}}\right) \tag{4.31}$$

where σ_0 is the yield stress at zero plastic strain, Q_{inf} and b are material hardening parameters, and $\dot{\bar{\varepsilon}}_{pl}$ is the equivalent plastic strain. The kinematic hardening law is obtained from [34]:

$$\dot{\alpha} = \sum_i [C_i \frac{1}{\sigma^0}(\sigma - \alpha)\dot{\bar{\varepsilon}}_{pl} - \gamma_i \alpha \dot{\bar{\varepsilon}}_{pl}] \tag{4.32}$$

where $\dot{\alpha}$ is the new yield surface due to kinematic hardening, C_i and γ_i are materials parameter, σ and α are the stress and back stress tensors (determining the translation of the yield surface). A Lemaitre-Chaboche hardening model takes into account both isotropic and kinematic hardening.

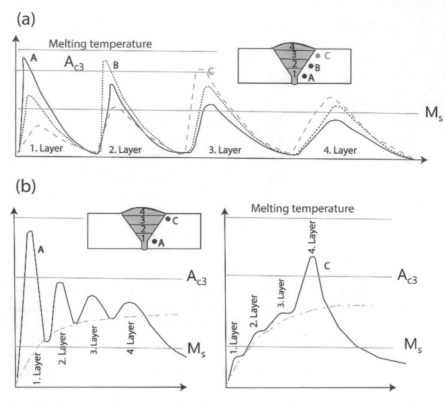

Fig. 4.18 Thermal cycles in some point resulted from multi-pass welding for **a** a long weld line, and **b** a short weld line [35]

In multi-layer welding, creep and annealing effects can occur, which are temperature-dependent phenomena. The former causes stress relief and the latter causes microstructural change. It is reported that taking the annealing effect into account leads to more accurate results than what is achieved by taking into account creep [18]. In fact, during numerical simulation one can generally ignore viscoplastic behavior in cases where there is neither a multi-pass welding nor a stress relief heat treatment. Usually, plastic strain is eliminated by annealing which occurs above a specific annealing temperature and a practical methodology to consider the influence of annealing is to remove any prior strain hardening in a isotropic hardening model in which the yield surface returns to its initial state when the local temperature exceeds the annealing temperature [36]. For example, this temperature for austenitic stainless steel is 1000 °C. In the simulation of a multi-pass welding, considering isotropic hardening, both kinematic hardening and annealing can improve the accuracy of the simulation to predict the residual stress. However, a kinematic hardening model requires several parameters that must be experimentally determined and also leads to larger computational times. It is reported that during the simulation of austenitic stainless steel welding, neglecting kinematic hardening gives a more conservative

Fig. 4.19 Schematic representation of yield surface and stress–strain curve for different material behaviors. **a** Isotropic hardening. **b** Kinematic hardening. **c** Combined isotropic-kinematic hardening [37]. Adopted with permission from Elsevier

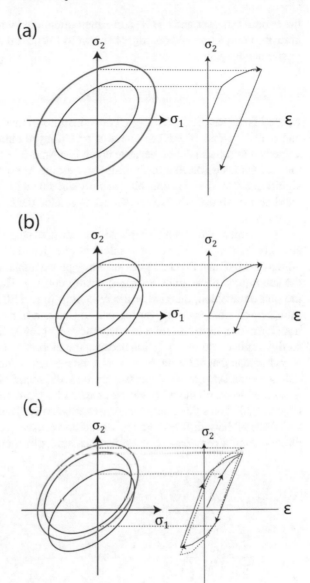

result of residual stress magnitude which in engineering applications is desirable [36].

In creep or visco-plastic constitutive models, the in-deformation is time dependent and in stress relieving process creep is almost always present. A creep process has three main stages (primary, secondary and tertiary). When simulating stress relief through creep in the welding, only the second stationary stage should be considered [38]. During a post-weld heat treatment, the process of creep reduces the residual stresses and can be performed uncoupled with the welding process, i.e. after the completion and cooling of the weld [39]. Norton's law describes well this phenomenon quantitatively [40]:

$$\varepsilon = \dot{C}_1 \sigma^{C_2} e^{-\frac{C_3}{T}} \tag{4.33}$$

where $\dot{\varepsilon}$ is the strain rate, C_1, C_2, C_3 are constants of material, 'T' is the temperature, and σ is the stress. The creep effect of reheating in multi-phase welding is mostly neglected in the simulation because creep is a time-dependent phenomenon and the thermal cycles associated to multi-layer welding are too short for creep to play a significant role. Creep is thus only usually considered in the simulation of the post-weld heat treatment which aims mainly to reduce the residual stresses present after welding.

The temperature at which a phase transition occurs (in the solid state) can also strongly influence the level of residual stress. This is especially true for steels in which the HAZ and FZ undergoes austenite phase transformation during heating and the subsequent phase transformation during cooling. The lower the temperature of the phase transition, the greater the reduction in residual stress. In general, phases transformed from austenite during cooling have a higher volume and if the phase transformation from austenite to ferrite occurs at lower temperature, there is necessarily a higher volume expansion because the thermal volume expansion of austenite is higher than that of ferrite. In general, a phase transformation during cooling establishes compressive residual stresses in the weld metal [41] and this residual stress is superimposed on the other stresses caused by rapid cooling (shrinkage stresses). Figure 4.20 shows this phenomenon schematically. The phase transformation and shrinkage all have different effects. Shrinkage causes a tensile stress in longitudinal direction and a compressive stress in transverse direction of the base material. The

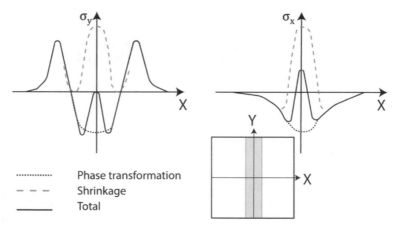

Fig. 4.20 Superposition of shrinkage stress and phase transformation stress in longitudinal and transverse direction [42]

thermal stresses caused by shrinkage are tensile in the weld zone while those caused by phase transformation are mainly compressive. The final shape of the residual stress distribution is then determined by the mechanism which is dominant (transformation, shrinkage, or rapid cooling). Furthermore, other factors such as the number of layers may also influence the stress profile.

In the case of steels which undergo martensitic transformation by cooling, the initial and final temperatures of martensitic transformation (M_s, M_f) as well as the temperature range for martensitic transformation $(M_s - M_f)$ have a large effect on the residual stress values [43]. This is due to the volumetric expansion during martensite transformation which compensates for the tensile residual stresses caused by shrinkage and the overall result is a compressive stress on the weld material. Low temperature transformation (LTT) steels are preferable due this. By cooling down to a temperature below the M_s the residual stress level starts to decrease until the temperature reaches M_f. At temperatures lower than M_f the residual stress bounces back and starts to increase. The relationship between the coefficient of thermal expansion (CTE), volume expansion strain (VES) and the temperature range of martensite formation (ΔT) is:

$$CTE = \frac{VES}{\Delta T} \tag{4.34}$$

As the martensite finish temperature in the weld decreases, more residual compressive stress is induced, which is desirable. During a simulation process, the volumetric change due to martensitic transformation can be represented by variation of thermal expansion coefficient with temperature, for two kinds of steels with two different transformation temperatures (LTT1 with $M_s = 245\,°C$, $M_f = 100\,°C$ and LTT2 with $M_s = 225\,°C$, $M_f < 0\,°C$). Here, a negative CTE can represent the volume increase due to martensite transformation. The simulation results show that by increasing the VES the compressive residual stress increases, as shown in Fig. 4.21. This also

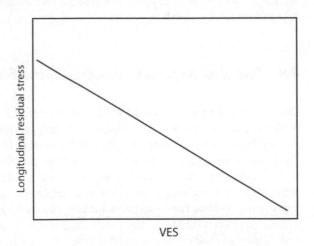

Fig. 4.21 Relationship between longitudinal residual stress and volume expansion strain (VES)

demonstrates how important it is to choose a filler material with a lower martensite finish temperature (for example LTT2 compared to LTT1), allowing to build up residual compressive stress.

In multi-pass welding, in which LTT materials were used, reheating of each layer by the subsequent layers can enhance the compressive residual stress if the reheat temperature exceeds the austenite transformation. Otherwise, a residual tensile stress would arise in each layer under the influence of the heat of the top layer [44]. This residual stress is very high, driven by the high yield strength of martensite, and can degrade the performance of the component, especially if the HAZ is martensitic as well. An alternative method to reduce residual stress, if the material volume expansion cannot change by martensite formation, is to use a different filler material with a lower yield strength, such as Ni-based filler materials [45]. In this way, although the residual stress adds up during multi-phase welding, it cannot increase in excess because it cannot exceed the nickel's low yield strength.

4.7 Weld Sequence

A fundamental action to reduce the residual stress is to properly design the welding sequence. Generally, a weld path can be sub-divided into a set of partial seams and therefore there are several possible combinations of weld sequences that can be simulated to calculate the resultant residual stress or the distortion level. As the number of sub-welds under consideration increases, so do the possible strategies. If one chooses to divide a path into n sub-welds, the total number of possible combinations would be equal to $2^n \times n!$. For a path with 4 sub-welds there is 384 possible strategies. In the case of spot welding, for an assembly consisting of N_w welds there is $N_w!$ Possible permutations. Running such a large number of simulations is very time-consuming and therefore some specific approaches must be followed to reduce the workload without deteriorating the accuracy of the results. In Chap. 5 some of these approaches are explained.

4.8 Technical Aspects of Residual Stress Simulation

One classical technique used to reduce computational time is to adopt a 2D FE model at the cost of a loss in complexity. In such simplification, a balance should be established between computing time and the detail being captured by the model. Of course, some geometries do not allow for this simplification and require the use of 3D models, such as the case of pipe grith weld in which the residual stresses are rotationally non-symmetric [9]. Another common process is to perform a thermal analysis divided into two main parts: heating and cooling. The results of the thermal history can then be loaded in the mechanical model to predict the residual stresses and distortion. Furthermore, instead of using the results of a moving heat source in

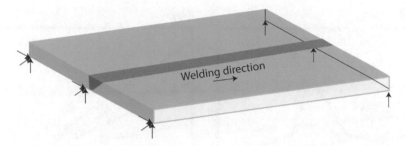

Fig. 4.22 The constrains in simulation of an aluminum alloy welded in butt configuration

the mechanical analysis, an instantaneous heat source can be adopted, although the mesh size is critical for the accuracy of the result [46]. Another strategy that is used to reduce the time it takes a CPU to compute the mechanical model is to limit the load to areas beyond which the heat contribution is negligible [47].

When a filler material is added to the weld, two techniques are utilized to account for the added material [48]: the quiet element (dummy) technique and the element death-rebirth technique. In the dummy technique the elements are active all the time and are initially assigned with dummy material properties. The actual weld material properties are reassigned only after the elements start to cool down. In the element rebirth technique, accurate material properties are initially assigned to elements, but they are inactivated before the corresponding part (the weld material) is deposited. These two techniques are known to be equally effective in large scale problems, with no considerable differences. However, it is advisable to examine the performance of both techniques with regards to the temperature distribution and weld nugget shape.

The constraints on the structures should also be carefully considered. Figure 4.22 shows the constraints in a simulation of an aluminum alloy welded in butt configuration. One edge of the lower surface was fully constrained while the other edge was constrained only along the plate thickness direction. Usually, the out of plane displacement is regarded as the distortion criterion. Constraints play a very important role on the residual stress and the final distortion of the component.

For the determination of accurate results it is necessary to consider the variation of both mechanical and physical properties of material with respect to temperature [49]. The physical properties are considered in thermal analysis and the mechanical properties are considered in mechanical analysis. Figure 4.23 shows an example of variation of materials properties with respect to temperature in a steel.

The element types adopted during the thermal analysis and structural analysis are almost always different. For instance, the literature shows examples where the element types chosen for thermal and mechanical analyses were DC3D20 (a 3D, 20-node quadratic isoparametric element for heat transfer) and C3D8R (a 3D, 8-node linear isoparametric element with reduced integration), respectively [48]. Triangular elements are generally stiff and are not well suited for elastic–plastic analysis. Meshing in a coupled thermo mechanical analysis requires elements that possesses

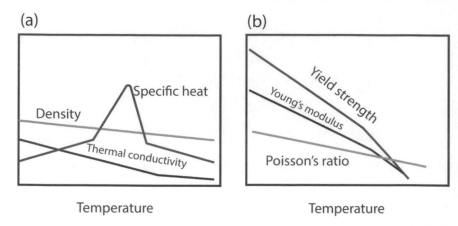

Fig. 4.23 Variation of **a** thermal properties, and **b** mechanical properties of a steel with temperature

both thermal and mechanical degree of freedom. Usually, additional software packages (such as Hypermesh or Femap) are used to mesh complex shapes in order to be further used by simulation software such as Abaqus, Sysweld and so on. Furthermore, in cases where large strains are present, it is fundamental to reduce the step size or mesh size, greatly lengthening the simulation time.

In welding process models, the peak temperature at various locations of the mesh will vary greatly. In the case of steel alloys this causes different microstructures to appear in the HAZ and every microstructure experiences a different thermal strain during cooling. Therefore, a fine mesh is needed in this region in order to take into account these transformations. To reduce computational costs associated to this detailed mesh, a technique named Reduced Geometry can be used which is based on the hypothesis that the temporal and spatial variation of temperature or other parameters is significant only around the welding region [50]. In this regard, only elements in the weld region need to be fine and, in the regions, far from the weld larger elements can be used in order to shorten the simulation time. The finer element size is mandatory to capture the steep gradients of temperature around the weld zone. Furthermore, there is no interest to record the variation of temperature far from the weld zone. Figure 4.24 shows the reduced and standard geometry.

The transient state of welding is decisive on establishment of the steady state condition. The rate of change and the local gradient in transient state are quite high and therefore in the simulation the mesh size and time step must be refined. This lengthens the time of analysis, especially when the metallurgical changes are considered in simulation.

The material characteristics which are needed for the simulation of residual stress are the thermal expansion coefficient, elastic modulus, Poisson's ratio, and yield limit. While the thermal expansion coefficient is temperature dependent, these properties are usually only needed only at low temperatures where the flow stress is high. Simple experiments can be performed to calibrate the simulation and obtain accurate results.

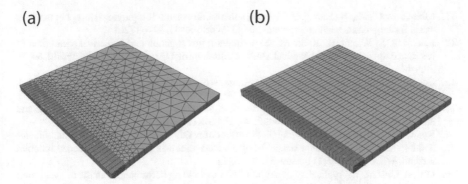

(a)   (b)

Fig. 4.24 Mesh of the computational model made with **a** reduced geometry, and **b** standard geometry

In this procedure, local properties of the weld in the model are modified and the simulation is repeated until an acceptable match is obtained between the simulation result and the experiment. In industrial applications, the modified properties often do not necessarily closely match the actual properties of the material and changes can be made, for example, to the yield point or other material properties.

References

1. Gong, K., et al.: Effect of dissolved oxygen concentration on stress corrosion cracking behavior of pre-corroded X100 steel base metal and the welded joint in wet–dry cycle conditions. J. Nat. Gas Sci. Eng.**77**, 103264 (2020)
2. Javadi, Y., et al.: Investigating the effect of residual stress on hydrogen cracking in multi-pass robotic welding through process compatible non-destructive testing. J. Manuf. Process. **63**, 80–87 (2021)
3. Li, L., et al.: Experimental and numerical investigation of effects of residual stress and its release on fatigue strength of typical FPSO-unit welded joint. Ocean Eng. **196**, 106858 (2020)
4. Tietz, H.-D.: Grundlagen der Eigenspannungen: Entstehung in Metallen, Hochpolymeren und silikatischen Werkstoffen, Meßtechnik und Bewertung; mit 12 Tab. 1982: Deutscher Verlag für Grundstoffindustrie
5. Hildebrand, J.: *Numerische Schweißsimulation-Bestimmung von Temperatur, Gefüge und Eigenspannung an Schweißverbindungen aus Stahl-und Glaswerkstoffen* (2009)
6. Kannengießer, T.: Untersuchungen zur Entstehung schweißbedingter Spannungen und Verformungen bei variablen Einspannbedingungen im Bauteilschweißversuch. Shaker (2000)
7. Hensel, J., Nitschke-Pagel, T., Dilger, K.: On the effects of austenite phase transformation on welding residual stresses in non-load carrying longitudinal welds. Weld. World **59**(2), 179–190 (2015)
8. Daichendt, S.: Relaxation strahlbedingter Eigenspannungen unter kombiniert thermisch-mechanischer Beanspruchung in Laserstrahl-Schweißverbindungen einer AlSiMgCu-Knetlegierung. Universität Bremen (2011)
9. Mirzaee-Sisan, A., Wu, G.: Residual stress in pipeline girth welds-A review of recent data and modelling. Int. J. Press. Vessels Pip. **169**, 142–152 (2019)
10. Radaj, D.: Heat effects of welding: temperature field, residual stress, distortion. Springer Science & Business Media (2012)

11. Murakawa, H., Ma, N., Huang, H.: Iterative substructure method employing concept of inherent strain for large-scale welding problems. Weld. World **59**(1), 53–63 (2015)
12. Pakkanen, J., Vallant, R., Kičin, M.: Experimental investigation and numerical simulation of resistance spot welding for residual stress evaluation of DP1000 steel. Weld. World **60**(3), 393–402 (2016)
13. Lu, Y., et al.: Numerical simulation of residual stresses in aluminum alloy welded joints. J. Manuf. Process. **50**, 380–393 (2020)
14. Huang. H., et al.: Toward large-scale simulation of residual stress and distortion in wire and arc additive manufacturing. Add. Manuf. **34**, 101248 (2020)
15. Kyaw, P.M., et al.: Numerical study on the effect of residual stresses on stress intensity factor and fatigue life for a surface-cracked T-butt welded joint using numerical influence function method. Weld. World **65**(11), 2169–2184 (2021)
16. Ma, N., Umezu, Y.: Application of explicit FEM to welding deformation: analysis. Weld. Int.: Trans. Worlds Weld. Press **23**(1), 1–8 (2009)
17. Ikushima, K., Shibahara, M.: Prediction of residual stresses in multi-pass welded joint using idealized explicit FEM accelerated by a GPU. Comput. Mater. Sci. **93**, 62–67 (2014)
18. Maekawa, A., et al.: Fast three-dimensional multipass welding simulation using an iterative substructure method. J. Mater. Process. Technol. **215**, 30–41 (2015)
19. Tremarin, R.C., Pravia, Z.M.C.: Analysis of the influence of residual stress on fatigue life of welded joints. Latin Am. J. Solids Struct. **17** (2020)
20. Islam, M., et al.: Simulation-based numerical optimization of arc welding process for reduced distortion in welded structures. Finite Elem. Anal. Des. **84**, 54–64 (2014)
21. Radaj, D.: Eigenspannungen und Verzug beim Schweissen: Rechen-und Messverfahren, Fachbuchreihe Schweisstechnik, vol. 143. Verlag für Schweißen und Verwandte Verfahren, DVS-Verlag, Düsseldorf (2002)
22. Murakawa, H., Deng, D., Ma, N.: Concept of inherent strain, inherent stress, inherent deformation and inherent force for prediction of welding distortion and residual stress. Trans. JWRI **39**(2), 103–105 (2010)
23. Wu, C., Wang, C., Kim, J.-W.: Welding distortion prediction for multi-seam welded pipe structures using equivalent thermal strain method. J. Weld. Join. **39**(4), 435–444 (2021)
24. Honaryar, A., et al.: Numerical and experimental investigations of outside corner joints welding deformation of an aluminum autonomous catamaran vehicle by inherent strain/deformation FE analysis. Ocean Eng. **200**, 106976 (2020)
25. Wang, J., et al.: Numerical prediction and mitigation of out-of-plane welding distortion in ship panel structure by elastic FE analysis. Mar. Struct. **34**, 135–155 (2013)
26. Murakawa, H., et al.: Applications of inherent strain and interface element to simulation of welding deformation in thin plate structures. Comput. Mater. Sci. **51**(1), 43–52 (2012)
27. Kim, T.-J., Jang, B.-S., Kang, S.-W.: Welding deformation analysis based on improved equivalent strain method considering the effect of temperature gradients. Int. J. Naval Archit. Ocean Eng. **7**(1), 157–173 (2015)
28. Yi, B., Wang, J.: Mechanism clarification of mitigating welding induced buckling by transient thermal tensioning based on inherent strain theory. J. Manuf. Process. **68**, 1280–1294 (2021)
29. Ma, N., et al.: Inherent strain method for residual stress measurement and welding distortion prediction. In International Conference on Offshore Mechanics and Arctic Engineering. American Society of Mechanical Engineers (2016)
30. Ha, Y.S., Cho, S.H., Jang, T.W.: Development of welding distortion analysis method using residual strain as boundary condition. In Materials Science Forum. Trans Tech Publ. (2008)
31. Wu, C., Kim, J.-W.: Numerical prediction of deformation in thin-plate welded joints using equivalent thermal strain method. Thin-Walled Struct. **157**, 107033 (2020)
32. Smith, M.C., Smith, A.C.: Advances in weld residual stress prediction: a review of the NeT TG4 simulation round robins part 2, mechanical analyses. Int. J. Press. Vessels Pip. **164**, 130–165 (2018)
33. Argyris, J., Szimmat, J.: Finite element analysis of arc-welding processes. In: International Conference on Numerical Methods in Problems, vol. 3 (1983)

34. Muránsky, O., et al.: Numerical analysis of retained residual stresses in C (T) specimen extracted from a multi-pass austenitic weld and their effect on crack growth. Eng. Fract. Mech. **126**, 40–53 (2014)
35. Rykalin, N., Fritzsche, C.: Berechnung der wärmevorgänge beim schweissen: aus dem Russ. Verlag Technik (1957)
36. Deng, D., et al.: Influence of material model on prediction accuracy of welding residual stress in an austenitic stainless steel multi-pass butt-welded joint. J. Mater. Eng. Perform. **26**(4), 1494–1505 (2017)
37. Teimouri, R., Amini, S., Guagliano, M.: Analytical modeling of ultrasonic surface burnishing process: evaluation of residual stress field distribution and strip deflection. Mater. Sci. Eng., A **747**, 208–224 (2019)
38. Ren, S., et al.: Finite element analysis of residual stress in 2.25 Cr-1Mo steel pipe during welding and heat treatment process. J. Manuf. Proc. **47**:110–118 (2019)
39. Deshpande, A.A., et al.: Combined butt joint welding and post weld heat treatment simulation using SYSWELD and ABAQUS. Proc. Instit. Mech. Eng. Part L: J. Mater.: Des. Appl. **225**(1), 1–10 (2011)
40. Tan, L., Zhang, J.: Effect of pass increasing on interpass stress evolution in nuclear rotor pipes. Sci. Technol. Weld. Joining **21**(7), 585–591 (2016)
41. Neubert, S.: *Simulationsgestützte Einflussanalyse der Eigenspannungs-und Verzugsausbildung beim Schweißen mit artgleichen und nichtartgleichen Zusatzwerkstoffen.* Technische Universitaet Berlin (Germany) (2018)
42. Wohlfahrt, H., Nitschke-Pagel, T., Kaßner, M.: Schweißbedingte Eigenspannungen-Entstehung und Erfassung. Auswirkung und Bewertung. DVS BERICHTE **187**, 6–13 (1997)
43. Feng, Z., et al.: Transformation temperatures, mechanical properties and residual stress of two low-transformation-temperature weld metals. Sci. Technol. Weld. Join. **26**(2), 144–152 (2021)
44. Feng, Z., et al.: Investigation of the residual stress in a multi-pass T-welded joint using low transformation temperature welding wire. Materials **14**(2), 325 (2021)
45. Guo, Q., et al.: Influence of filler metal on residual stress in multi-pass repair welding of thick P91 steel pipe. Int. J. Adv. Manuf. Technol. **110**(11), 2977–2989 (2020)
46. Pu, X., et al.: Simulating welding residual stress and deformation in a multi-pass butt-welded joint considering balance between computing time and prediction accuracy. Int. J. Adv. Manuf. Technol. **93**(5), 2215–2226 (2017)
47. Tchoumi, T., Peyraut, F., Bolot, R.: Influence of the welding speed on the distortion of thin stainless steel plates—Numerical and experimental investigations in the framework of the food industry machines. J. Mater. Process. Technol. **229**, 216–229 (2016)
48. Ahn, J., et al.: Determination of residual stresses in fibre laser welded AA2024-T3 T-joints by numerical simulation and neutron diffraction. Mater. Sci. Eng., A **712**, 685–703 (2018)
49. Yu, H., et al.: Numerical simulation optimization for laser welding parameter of 5A90 Al-Li alloy and its experiment verification. J. Adhes. Sci. Technol. **33**(2), 137–155 (2019)
50. Farias, R., Teixeira, P., Vilarinho, L.: An efficient computational approach for heat source optimization in numerical simulations of arc welding processes. J. Constr. Steel Res. **176**, 106382 (2021)

Chapter 5
Data Based Simulation

Abstract Although the use of computational methods has significantly reduced the cost of weld design and optimization processes, eliminating the number of experiments needed, there are still important limitations on the capabilities of these processes. In many cases, multiple modelling runs may be needed, greatly increasing the length of the design process. However, multiple strategies have been developed to reduce the number of runs needed, greatly increasing the effectiveness of using models in welded joint design workflows. Calibration procedures are also fundamental for modelling welded joints, being essential to ensure that the model correlates well with experimental data and produces valid results. Optimization strategies can be used also to calibrate the simulation parameters as well and some of the most important of these approaches are described in detail in this chapter.

5.1 Introduction

Typically, the use of data analysis in process simulation has two main aims. One is to optimize the simulation parameters to accurately calibrate the model, attaining reliable results while running as few simulations as possible. The other goal is to optimize the process parameters, achieved by using an already calibrated simulation model. In this way, instead of carrying out multiple practical experiments, one can rely on a properly validated simulation to freely optimize the process parameters. Still, although simulation opens the door to situations where experimental activities may be very limited or even no longer required, running multiple simulations can still be a very time-consuming process and appropriate strategies should be used to reduce the number of models and simulations employed. This chapter explains the methods that are used to optimize either the simulation parameters or the process based on the simulation.

5.2   Optimization of Simulation Parameters

Models used for welding simulation purposes, such as heat source models, usually rely on a large number of parameters. The purpose of an optimization process is to find the most accurate values for these parameters, which must be calibrated in relation to a defined output or a combination of outputs. In one approach, a relationship between the model parameters and the output is established by running multiple simulations over a predefined parameter field.

One key challenge associated to optimizing welding simulation problems relies on how to obtain the exact parameters of the heat source model. The input parameters are defined depending on the heat source model and consist of geometric values of the heat source and of the heat quantity. The outputs can be the geometrical dimensions of the weld pool and/or the time–temperature history at some specific locations on the joint. The melt zone boundary (the shape of the weld pool) is reported to be a more reliable output to consider when calibrating the heat source parameters [1].

The Goldak model used for welding heat sources has 5 variables (a, b, c_f, c_r, Q). Welding pool characteristics such as fusion width W, penetration depth D, and the peak temperature T_p are used to calibrate the heat source [2]. Both natural logarithmic and linear regression models can be used to correlate the welding pool characteristics to the input variables, as follows:

$$f_{log,i}\left(Q, a, b, c_f, c_r\right) = \alpha_i Q^{\beta 1_i} a^{\beta 2_i} b^{\beta 3_i} c_f^{\beta 4_i} c_r^{\beta 5_i} \tag{5.1}$$

$$f_{l,i}\left(Q, a, b, c_f, c_r\right) = \alpha_i + \beta 1_i Q + \beta 2_i a + \beta 3_i b + \beta 4_i c_f + \beta 5_i c_r \tag{5.2}$$

where $f_{log,i}\left(Q, a, b, c_f, c_r\right)$ and $f_{l,i}\left(Q, a, b, c_f, c_r\right)$ are the characteristic of the welding pool (W,D or T_p), subscript of log represents nature logarithm regression model and subscript l means linear regression model and i represents fusion width (mm), penetration depth (mm) and peak temperature (∘C) at each welding conditions. Q, a, b, c_f, c_r are the heat source parameters and ai, β1i, β2i, β3i, β4i,β5i are the coefficients of each parameter in the regression equations.

In another published approach, principal component analysis (PCA) is used to substitute the independent variables with a new variable. This new variable can be a linear combination of original variables [2]. In fact, the main reason behind the use of regression analysis is to understand the contribution of each variable on the output. Sensitivity analysis can also be performed to assess the effect of input parameters. The range for each parameter needs to be precisely defined in order to allow for proper design of the input variables. Usually, a full analysis by full factorial design is highly resource intensive and alternative strategies are thus needed to reduce computational time. For a main factor analysis, a two-level fractional factorial design is a simple way to analyse the multi-factor influence [3]. In fractional factorial design, the interaction between the parameters is scarified for a smaller number of simulations. For levels greater than 2, the response surface method takes less time than the fractional factorial analysis. The response surface methodology can then be used to optimize the unknown coefficients, especially when their number is small [4].

Some simulations require the model to be calibrated by obtaining the optimal variable, which is usually a material property. For example, in the simulation of resistance spot welding, which is commonly performed with commercially available software, the contact electrical resistance between the two materials is unknown. The proper value for this parameter can be obtained via comparison between the weld nugget diameter obtained from simulation and that measured experimentally [5].

5.3 Simulation Supported Process Optimization

Every welding process is governed by a set of key variables that influence the welding properties. Before the development of process simulation software, numerical methods and the availability of powerful computers to run it, optimization of welding processes was solely based on experimentation activities [6]. Nowadays, welding processes are optimized in various ways, which can include only simulation activities or combine experimental and numerical stages. Still, with the growth and widespread availably of high-performance computing and the development of more capable commercial software, the use of simulation-based optimization has greatly expanded. However, simulation-based methods still require some experimentation to calibrate the simulation and verify the optimal state, a process which is explained in more detail later in this section. A simulation of the process can be carried out to optimize the parameters with regard to the desired properties of the weld [7]. Although it reduces the large costs associated to experimentation, the number of simulations that can be performed for optimization is limited because of the complexity associated to these models. Therefore, strategies must be implemented to reduce the number of simulations required and to save time. Some of these strategies are discussed below.

The first step in an optimization process consists in the determination of the parameters of the welding process which require optimization. This also requires the determination of the allowable range for each parameter, which depends on diverse factors, such as the material type, thickness, and joint configuration. Oftentimes, economic aspects are also highly important to determine the limits for process parameters. For example, a minimum welding speed can be specified in order to reduce welding costs [8]. Commonly used variables for conventional welding processes can be found in Table 5.1.

Following the determination of the process parameters and their allowable ranges, one must determine the output parameters and their target values. For example, one may seek to minimize the residual stress or distortion in a specific location. In Fig. 5.1, are some examples of the outputs that should be optimized, in which the bead width (BW) and the penetration depth (DP) are to be minimized and maximized, respectively [9].

In friction stir welding of aluminium alloys, temperature is a key factor in material flow and therefore any defects related to material flow during welding are considered

Table 5.1 Welding variables for various welding processes to be defined for optimization

Welding process	Welding parameters			
GMAW	Wire feed rate (current)	Voltage	Welding speed	
TIG	Current	Welding speed		
Laser welding	Laser power	Working distance	Focal position	Welding speed
FSW	Rotation speed	Welding speed	Normal load	

Fig. 5.1 The weld bead profile and the targeted dimensions

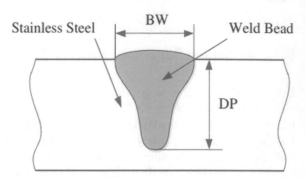

to be temperature dependent [10]. If the temperature is too high, the excess material flow leads to the formation of flash and if the temperature is too low, a deficit in material flow leads to lack of consolidation and the formation of wormhole defects. The maximum peak temperature can be considered as a good criterion for predicting weld defects. For example, the maximum permissible temperature achieved during the FSW of aluminium alloys (T_{max}) in order to obtain a defect-free weld seam is determined according to the following equation [11]:

$$\frac{T_{max}}{T_s} = 1.344 - 5.917 \times 10^{-4} T_s \tag{5.3}$$

where T_s is the solidus temperature.

In the case of the FSW process, the peak temperature is correlated with the heat generation and the heat generation can be correlated to the welding parameters (welding speed and rotational speed) with adequate analytical models. In this case, the optimization of the process parameters in order to obtain a defect-free weld can be performed bases on analytical formula.

Another aim of optimization methods is to minimize the geometrical changes in an assembly consisting of spot welds. In the automotive industry, spot welding for the body-in-white assembly process consists of two stages. The first step is spot welding while clamping the parts, which is designed to lock the geometry in place after it is released from the clamp. The geometrical change caused by the springback after clamping is influenced by the sequence of the spot welds. This geometrical change must be minimized by optimizing the spot-welding sequence. The geometrical change can be defined as the root mean square value of the deviation of each node from its nominal position (Q) and is derived from [12]:

$$q_i = q_u^i - q_{nom}^i \tag{5.4}$$

$$Q = \sqrt{\frac{1}{N} \sum_{i=1}^{N} q_i^2} \tag{5.5}$$

where q_i is the deviation of node I from its nominal position, q_u is the position of the node after welding and spring-back, q_{nom} is the nominal position of the node, and N is the number of nodes in the assembly.

In addition to optimizing a specific parameter, one may also need to consider the use of penalty terms to ensure that all aspects of the weld are accounted for. For example, if one optimizes solely to minimize distortion, this might lead to welds with insufficient penetration and poor strength. One way to solve this problem is to define an upper/lower bond for welding variables to ensures that the weld possesses exhibit full penetration. Another way is to define a penalty term that forces penetration to be within a desirable range.

The second step is to construct a FEM model to simulate the process. This process itself may require an optimization process as described in the previous section.

In the third step, a model is developed to correlate the input and output parameters. Various models can be used to achieve this, such as regression, polynomial response surface [13], artificial neural network (ANN) [14], radial basis function (RBF) [15], and Kriging [9].

The fourth step is to use the obtained model in the third step for optimization. Various optimization methods, such as simulated annealing algorithm [16] and nondominated sorting genetic algorithm (NSGA-II) [9] can be used for optimization purposes. In this case, the optimal input parameters can be obtained, but the result must still be validated against experimental data.

In some cases, one may entirely avoid the third stage if the evaluation of the population (when using genetic algorithm) is performed via simulation. In Fig. 5.2 the stages of genetic algorithm are shown. The population evaluation can be performed either by the verified simulation (obtained in the second step) or by the model which correlates the input parameters with output (obtained in the third step).

A study was performed to optimize the laser transition welding parameters with respect to the temperature during welding of thermoplastic material [17]. In this case, low temperature causes insufficient heating and melting and high temperature fosters crystallinity of the material which deteriorates the mechanical properties. The purpose of the optimization was to ensure that the temperature remains within temperature window of the thermoplastic. A penalty was also considered to account for the cost of the welding process.

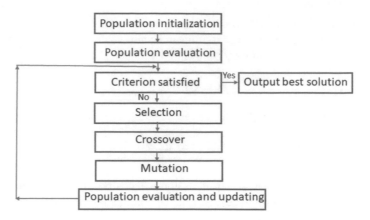

Fig. 5.2 Description of genetic algorithm

Another study investigated the influence of laser welding parameters on the peak temperature and the aspect ratio when welding Hastelloy, a type of nickel based superalloy [18]. First, the process was optimized by running several experiments designed based on Taguchi method. A regression analysis was developed to predict the responses and the optimization was performed by computing the signal to noise ratio. ANN was also used to predict the response with respect to the welding parameters.

In another study, the aim was to optimize the FSW process with regards to the temperature and economical aspects [8]. The peak temperature should be kept at 443 °C and the welding speed limited within a certain range in order to minimize welding costs by maximizing the welding speed and at the same time preventing the formation of defects at high travel speeds. The approach used for this optimization consisted of the stages shown in Fig. 5.3. First, a numerical model based on finite difference method was developed for thermal analysis. This model was calibrated and verified by experimental data. Then, a response surface for the peak temperature in relation to the welding parameters was derived. Thereafter, the welding speed was formulated as a function of temperature and rotation speed. Finally, the welding speed was maximized by using the exterior penalty method, considering constraints on the welding speed, the rotation speed, and the temperature. The constraint on welding speed (V_a) was intended to both avoid any weld defect (upper limit) and lower the cost of welding (lower limit). The constrain on the rotation speed was considered based on the capability of the FSW equipment.

Optimization methods are also used to obtain the optimal welding sequence in order to control the distortion caused by welding. One strategy to reduce the distortion is to use sub-passes with carefully planned sequences. The displacement caused by a sub-pass weld depends on its direction and location as well as the combined effect of all previous weld passes. By increasing the number of sub-passes, the possible welding sequences increase and therefore there would be a huge number of combinations (for n sub-passes there is $2^n \times n!$ combinations). Even with simulation,

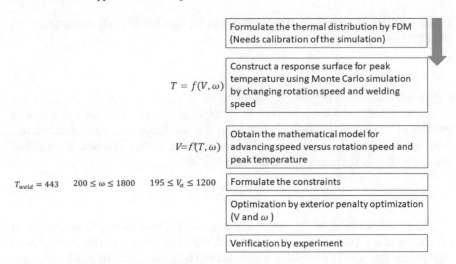

Fig. 5.3. Optimization stages of the speed of the FSW process with penalties on weld defect, equipment capability, economy, and temperature

it is not possible to run through all combinations, since the computing time needed for calculating the residual stresses for each combination is considerable. A surrogate model can be used to reduce the computation time significantly and at the same time to approximately predict the results for all combinations [19]. A surrogate model is a cheap model (in terms of computation time) that can represent a complex model [20]. These models are also very specific in its capabilities, further reducing computational time. The stages for making a surrogate model are [21]:

1. Creating a design of experiments for initial runs;
2. Evaluate the objective functions over those initial runs, either experimentally or by simulation;
3. Train a surrogate model;
4. Carry out the optimization using the trained surrogate model;
5. Evaluate the optimized design by the methods used in the Step 2;
6. Compare the result from the surrogate model with experiment;
7. If the results are satisfactory, advance to Step 9;
8. Add this design to the existing design and return to Step 3;
9. Stop and finalize the optimization.

Step 3 helps reduce computational time, as only a limited number of runs (simulation or experiment) are required. The surrogate model is based on this fact that only the initial sub-passes contribute to the main distortion during welding, as the next passes encounter a high stiffness scenario due to the existence of previous welds.

In an assembly with several spot welds the number of possible mutations is $N_w!$. In spot welding only the initial weld points play a critical role on the geometry. Based on this, only the sequence in the initial weld points is critical for determining the whole sequence. This helps to reduce the number of samples by taking only the "s" initial numbers into consideration. In this way the number of evaluations decreases to [12]:

$$\frac{N_w!}{(N_w - s)!} \tag{5.6}$$

As "s" decreases, a lower number of simulations is needed but there is a critical number of "s" below which the accuracy of the prediction becomes unacceptable. The correlation between the root mean square of distortion (Q) and the welding sequence (ξ) was determined with the help of RBF:

$$Q = f(\xi) \tag{5.7}$$

The minimum value of "s" which can accurately represent the distortion (Q) in all combinations (where $N_w = 7$) was equal to 3. The process of this evaluation is depicted in Fig. 5.4. The trends of variation of Q with respect to all possible combinations are similar for s = 3–6. By the obtained surrogate model with s = 3) only 4 percent of all possible weld sequences can represent all the possible weld sequences. Once this procedure for finding the minimum values of s has been performed for one assembly, it is safe to apply it to other assemblies. A RBF NN model can be trained for each assembly to predict Q with all permutations (all welding sequences).

The advantage of using the surrogate method over GA is that many evaluators can be used concurrently to reduce optimization time. With the surrogate method, sampling is carried out first and the responses for each ample can be evaluated independently by using multiple evaluators such as Computer Aided Tolerancing (CAT). Thus, if one can increase the number of evaluators, it becomes possible to reduce the computing time, since each sample can be evaluated independently of others. In GA, each evaluation is dependent on the previous one, as the process of sampling is sequential, and therefore adding the number of evaluators does not bring any benefit regarding the minimization of computational time [12].

In an attempt to minimize the distortion in a structure composed of lap joint made by arc welding process, a simulation-based optimization based on genetic algorithm was used [22]. The maximum distortion in the elements ($max(D_i)$) was considered

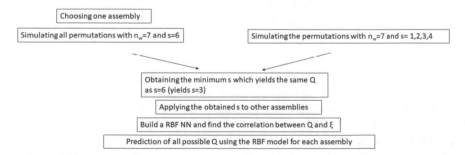

Fig. 5.4. The process of optimization of distortion (Q) with surrogate method

as the objective function for minimization $(F(x))$.

$$F(X) = \max(D_i) \tag{5.8}$$

$$D_i = \sqrt{(d_x)_i^2 + (d_y)_i^2 + (d_z)_i^2}, i = 1, 2, 3 \ldots .n \tag{5.9}$$

In which $(d_x)_i$ is the displacement of element i with relative to its initial location in the direction x, and N is number of elements in the structure. The input variables were defined as welding speed, arc voltage, current and welding direction. A numerical number was assigned for each possible welding direction. A penalty term was defined as the number of nodes whose maximum temperature is lower than the melting temperature in order to avoid defects in the welds. The optimization was performed by GA whose flowchart was shown in Fig. 5.2.

A rule-based genetic algorithm was used to obtain the optimum spot welding sequence to obtain the optimum geometrical quality [23]. In a standard genetic algorithm, a random individual is selected to initiate the optimization. A large initial population is needed to explore the searching space which is too time consuming to do with an iterative FEA simulation. In rule-based genetic algorithm the rules for spot weld sequencing replace the random position of initial individuals used in genetic algorithm which considerably reduces the computational time.

References

1. Goldak, J.A., Akhlaghi, M.: Computational welding mechanics. Springer Science & Business Media (2005)
2. Jia, X., et al.: A new method to estimate heat source parameters in gas metal arc welding simulation process. Fusion Eng. Des. **89**(1), 40–48 (2014)
3. George, E., et al.: Statistics for experimenters: design, innovation, and discovery, vol. 2. Wiley New York, NY, USA, (2005)
4. Tchoumi, T., Peyraut, F., Bolot, R.: Influence of the welding speed on the distortion of thin stainless steel plates—numerical and experimental investigations in the framework of the food industry machines. J. Mater. Process. Technol. **229**, 216–229 (2016)
5. Schulz, E., et al.: Short-pulse resistance spot welding of aluminum alloy 6016–T4—Part. Weld. J. **100**(3), 83S-92S (2021)
6. Benyounis, K., Olabi, A.-G.: Optimization of different welding processes using statistical and numerical approaches–a reference guide. Adv. Eng. Softw. **39**(6), 483–496 (2008)
7. Chen, F.F., et al.: Model-based parameter optimization for arc welding process simulation. Appl. Math. Model. **81**, 386–400 (2020)
8. Fraser, K.A., St-Georges, L., Kiss, L.I.: Optimization of friction stir welding tool advance speed via monte-carlo simulation of the friction stir welding process. Materials **7**(5), 3435–3452 (2014)
9. Jiang, P., et al.: Optimization of laser welding process parameters of stainless steel 316L using FEM, Kriging and NSGA-II. Adv. Eng. Softw. **99**, 147–160 (2016)
10. Arbegast, W.J.: A flow-partitioned deformation zone model for defect formation during friction stir welding. Scripta Mater. **58**(5), 372–376 (2008)

11. Qian, J., et al.: An analytical model to optimize rotation speed and travel speed of friction stir welding for defect-free joints. Scripta Mater. **68**(3–4), 175–178 (2013)
12. Tabar, R.S., Wärmefjord, K., Söderberg, R.: A new surrogate model–based method for individualized spot welding sequence optimization with respect to geometrical quality. Int. J. Adv. Manuf. Technol. **106**(5), 2333–2346 (2020)
13. Ruggiero, A., et al.: Weld-bead profile and costs optimisation of the CO_2 dissimilar laser welding process of low carbon steel and austenitic steel AISI316. Opt. Laser Technol. **43**(1), 82–90 (2011)
14. Olabi, A., et al.: An ANN and Taguchi algorithms integrated approach to the optimization of CO_2 laser welding. Adv. Eng. Softw. **37**(10), 643–648 (2006)
15. Ai, Y., et al.: Process modeling and parameter optimization using radial basis function neural network and genetic algorithm for laser welding of dissimilar materials. Appl. Phys. A **121**(2), 555–569 (2015)
16. Kolahan, F., Heidari, M.: A new approach for predicting and optimizing weld bead geometry in GMAW. Int. J. Mech. Syst. Sci. Eng. **2**(2), 138–142 (2010)
17. Labeas, G., Moraitis, G., Katsiropoulos, C.V.: Optimization of laser transmission welding process for thermoplastic composite parts using thermo-mechanical simulation. J. Compos. Mater. **44**(1), 113–130 (2010)
18. Bagchi, A., et al.: Numerical simulation and optimization in pulsed Nd: YAG laser welding of Hastelloy C-276 through Taguchi method and artificial neural network. Optik **146**, 80–89 (2017)
19. Asadi, M., Goldak, J.A.: Combinatorial optimization of weld sequence by using a surrogate model to mitigate a weld distortion. Int. J. Mech. Mater. Des. **7**(2), 123–139 (2011)
20. Jones, D.R., Schonlau, M., Welch, W.J.: Efficient global optimization of expensive black-box functions. J. Global Optim. **13**(4), 455–492 (1998)
21. Voutchkov, I., et al.: Weld sequence optimization: the use of surrogate models for solving sequential combinatorial problems. Comput. Methods Appl. Mech. Eng. **194**(30–33), 3535–3551 (2005)
22. Islam, M., et al.: Simulation-based numerical optimization of arc welding process for reduced distortion in welded structures. Finite Elem. Anal. Des. **84**, 54–64 (2014)
23. Sadeghi Tabar, R., et al.: A novel rule-based method for individualized spot welding sequence optimization with respect to geometrical quality. J. Manuf. Sci. Eng. **141**(11), 111013 (2019)

Printed in the United States
by Baker & Taylor Publisher Services